Gravity Energy Storage

Gravity Energy Storage

ASMAE BERRADA, PhD
School of Renewable Energy, LERMA
Université International de Rabat

KHALID LOUDIYI, PhD
School of Science and Engineering
Al Akhawayn University

ELSEVIER

Publisher: Joe Hayton
Acquisition Editor: Lisa Reading
Editorial Project Manager: Joanna Collett
Production Project Manager: Poulouse Joseph
Cover Designer: Alan Studholme

3251 Riverport Lane
St. Louis, Missouri 63043

Working together
to grow libraries in
developing countries

www.elsevier.com • www.bookaid.org

To my parents, sisters, Sofia and Zineb,
and my newborn nephew, Ghali,
for their inspiration.

Asmae Berrada

Preface

Gravity energy storage (GES) is an innovative energy storage technology that is attracting attention lately due to the many benefits it provides. GES systems have interesting characteristics and potential as it is based on the well-established principle of pumped hydro energy storage.

Research is needed to investigate the realization of such innovative storage technology. This is addressed in this book by combining technical, economic, and political approaches. Models have been developed to examine the different aspects of this storage system, which include design, profitability, feasibility, value, operation, behavior, development risk, and performance. In addition, political considerations and barriers have been discussed to complement these approaches.

Comprehensive analyses of GES and its applications are carried out throughout this book, providing the reader new insights, understanding, models, simulations, methods, along with several case studies. This book is divided into six chapters. Chapter 1 provides a general introduction and an overview about energy storage technologies. Novel energy storage systems based on the principle of pumped hydro energy storage are also discussed. Chapter 2 presents a technical design and material selection for GES components. In addition, a mathematical model is developed to optimally size an energy storage system based on an energy storage capacity while considering the different limitations of the system. It also analyses the feasibility of combining compressed air with gravity storage. The technical and economic aspects of this improved system are investigated in this chapter. In chapter 3, an economic study is performed to identify the levelized cost of energy for this system and compare it with other storage solutions.

It also values the potential of energy storage participating in energy and ancillary markets. In addition, the economic viability of energy storage in different grid applications has been examined. A risk study has been conducted to investigate whether the development of this innovative gravity storage system would be feasible. Chapter 4 discusses the services offered by energy storage to the utility grid. Furthermore, the economic profitability of GES operating in both small- and large-scale applications has been investigated. Chapter 5 delves into the dynamics of gravity storage mechanical and hydraulic components. Models have been developed to investigate the system performance and dynamic behavior. The simulation outcomes have been compared with experimental results to validate the proposed models. Chapter 6 addresses the challenges and barriers that hinder the development of novel energy storage systems.

This book concludes that GES is an attractive technology, which offers interesting technical and economic perspectives. If energy storage is to play an effective role as part of a low carbon transition, research and development of new storage system is necessary. It is our hope that the book would act as a trigger for the development of this novel energy storage system. Barriers facing the development of such systems need to be considered.

This book will be of interest to electrical and power engineers, practitioners in renewable energy storage systems, students, as well as researchers in energy storage development and implementation. It is hoped that this book will not only be of value to the aforementioned people but also to decision-makers, developers, and all interested people.

Acknowledgments

We would like to thank all the people who have contributed to the elaboration and the delivery of this work. Special thanks go to the following colleagues and graduate students for their efforts and dedicated time to help in the preparation of some sections:

- Safae El Kalai
- Mouna El Boukhari
- Wissal EL Fadil
- Iasse Fatima Zahra

We would like to acknowledge the support of our academic institutions and wish to thank the editor and Elsevier for their help in producing this book.

Special thanks go to our families and friends. They have been a source of motivation and support.

About the Authors

Asmae Berrada is a professor in the school of renewable energy at the international University of Rabat (UIR). She holds a PhD degree from the Faculty of Science at SMBA University in Fez. Her PhD work was prepared at Al Akhawayn University within the framework of the EUROSUNMED project (Euro-Mediterranean Cooperation on Research & Training in Sun Based Renewable Energies). She obtained a master degree of science in Sustainable Energy Management and a bachelor degree in Engineering and Management Science from Al Akhawayn University. Dr. Berrada started her career as a part-time faculty at Al Akhawayn University from 2014 to 2018. She has received a number of research awards and honors such as First Prize for Talent Award in Research and Innovation by Masen in 2018 and Best PhD thesis by SMBA. She has published a number of articles in peer-reviewed journals and has been a reviewer for several international journals known in the field of energy. Her main research focus is on energy systems and their modeling. She has actively been working in the area of energy storage.

Khalid Loudiyi Professor Khalid Loudiyi has been a faculty member of School of Science and Engineering at Al Akhawayn University in Ifrane (AUI), Morocco, since 1995. He joined AUI after spending 5 years (1089–1994) at Ibn Tofail University in Kenitra, Morocco, and 1 year as a visiting assistant professor at Oklahoma State University. Professor Loudiyi holds a PhD degree in physics from Oklahoma State University. During his academic career, he has established different academic programs at AUI and Ibn Tofail University. For the past 10 years, he has been working on different renewable energy projects. Among his contributions, at AUI, in this area are the setting up a Master of Science degree in "Sustainable Energy Management," and starting major renewable energy projects in energy storage, solar energy, wind energy, and energy efficiency.

Contents

CHAPTER 1

Energy Storage

INTRODUCTION

The development of the industrial sector throughout the years has led to a clear enhancement in living conditions. Accordingly, an increase in energy consumption has been witnessed [1]. As for now, many countries are focusing on developing the field of renewable energy (RE) and enhancing its control strategies and management [2]. Electricity generated from RE technologies is predicted to increase from only 10% in 2010 to 35% in 2050 [3]. The variability of energy produced by renewable sources represents one of the main challenges faced by the electric grid. As a matter of fact, one of the solutions to this issue is energy storage (ES). This solution has been proved to be an effective one for ensuring the integration of RE into the grid [4]. ES technologies have technical and economic advantages from the generation of energy to its supply [5]. Storage also balances demand and supply to avoid energy losses or power cuts. Hence, ES technologies compensate for any lags between generation and supply especially for RE generation and enhance the reliability of the utility grid [5].

ES systems are crucial in meeting power demand and in incorporating more environmentally friendly energy technologies. There are several benefits of using ES such as reducing energy costs, improving indoor air quality when using benign energy for heating or cooling, reducing energy consumption, increasing the operating flexibility, and last but not least, reducing operating and maintenance costs. Moreover, further advantages to the use of ES have been reported, which include reducing equipment size, increasing the operating efficiency and utilization of equipment, conserving and substituting fossil fuels by reducing their use, as well as decreasing GHG emissions.

ES systems have a great potential in increasing the efficiency and effectiveness of energy production systems by contributing to the substitution of fossil fuels. However, ES is difficult to understand and cannot be investigated without fully considering the whole electric value chain. To get the maximum benefit of a storage system, specific set of actions must be performed in the different energy sectors. Performance criteria of ES help in evaluating the performance and usefulness of the system to determine the possibility of developing more advanced systems. Evaluating ES systems should compromise the performance of a study about the potential impact of this system at different levels.

Nowadays, reliable supply of energy is a crucial aspect regarding how civilizations become more and more industrialized. For raw energy to be useful, it should be converted to what is called energy currencies. This is commonly done through heat release. As an example, industrial processes make use of steam for heating purposes. This later is created by transferring heat from burning fuel to water. Electricity is generated mainly with steam-driven turbogenerators; this steam is usually created using fossil fuels or nuclear energy as the source of heat. Demand on power in all of its forms is usually not steady, nor is the supply. For example, solar energy is not steady either for thermal or electric energy. In the past, the only solution to cope with the variability of supply and demand and to ensure the reliability of power supply was to convert large quantities of power to meet peak demand periods. This approach results in lower efficiencies and higher capital investments due to the fact that generating systems should be operating less than their full capacity most of the time.

Load management is a very important aspect because it helps in smoothing power demand and thus can sometimes leads to reducing capital investments. When ES is used, load management can also lead to reducing the capacities of power generation. Small capacity systems can still operate close to and around peak capacity regardless of the instantaneous demand. This is done through storing energy that was produced during low demand periods and which can be used later on during peak demand periods. Storage processes are not 100% efficient; therefore, some energy is lost while being stored. However, ES allows conserving fuels by using more abundant yet less flexible fuels such as coal and nuclear energy, instead of using less abundant fuels such as oil and natural gas. ES systems allow in some cases the reuse of heat waste from production processes.

Gravity Energy Storage https://doi.org/10.1016/B978-0-12-816717-5.00001-3

The use of ES is not only restricted to industries and production facilities; it can also be incorporated in residential and commercial sectors. It also plays an important role in integrating solar heating and cooling systems. It may also contribute to lowering peak demand loads resulting from conventional electric cooling and heating systems. For the transportation sector, in which the biggest share of vehicles are the gasoline-powered ones, ES systems will smoothen the transition toward electric vehicles resulting in lowering the demand for petroleum. ES systems may offer other services. For example, flywheels are able to smooth fluctuations; this service was known after the appearance of reciprocating engines in the 18th century.

In recent years, a renewed interest in the use of ES have been seen because of the increasing cost of fuel, the difficulties of obtaining capitals for the expansion of power generation capacities and the development of new storage systems. The value of energy for suppliers is determined by the demand and the production cost. As for costumers, the value of energy depends on its contribution to their personal comfort. Alternative energy production and consumption patterns are still being discussed; however, final decisions are most likely to keep being taken based on economic evaluations. When it comes to ES systems, decisions are basically made based on savings, except when regulations are imposed. Hence, economic potential and viability is one of the most important aspects that should be considered when commercializing ES systems.

There are two main types of ES systems. The first category deals with distributed ES, whereas the second one is about bulk ES. These systems compromise large storage capacities, which are used to supply energy to transmission application when the energy demand is high. The main types of bulk storage systems include pumped hydrostorage (PHS) and compressed air energy storage (CAES) [5,6]. Until now, the most common storage system used in the world is PHS with an installed capacity of about 145 GW [7]. This system represents 98% of the total installed storage capacity worldwide [8]. Moreover, PHS installed capacity is expected to increase in Europe by 20% in 2020 [9]. PHS is also seen as the most reliable ES technique [10]; it is considered a good solution for the integration of RE system into the grid. Cialis has shown that this storage technology is the best storage system to support high integration of wind farms [11]. Research has shown that PHES ensures the validity of RE resources [12]. Other projects have demonstrated the practicality of PHS for remote applications [13,14]. However,

PHS faces many challenges such as environmental concerns and geographic issues.

Both PHS and CAES technologies suffer from many issues and limitations. As a matter of fact, PHS has negative environmental impacts, requires high initial investment, and encounters difficulties in identifying appropriate implementation sites. CAES, on the other hand, has low energy efficiencies compared with other storage techniques such as batteries and is also constrained by the difficulty of determining appropriate underground reservoirs [15]. Even if PHS suffers from the abovementioned drawbacks, it is considered the most popular storage technology because of the many benefits it offers. For example, it is economically viable on the long run; it has large capacities, and long lifetime, and it is technically feasible as opposite to other storage technologies [16]. Researchers and industrials are currently developing technologies that are similar to PHS. A very interesting technology is the underground PHS. The working principle of this technology is similar to that of PHS except that the height difference is achieved in this system by digging underground. The lower reservoir can be either constructed or it can make use of already existing cavern or mine. The technical and economic constraints faced by this technology such as legal permissions, construction risks, and long return on investment periods are the main reasons why this technology is still not built [17]. As for now, many related projects are being developed. A conceptual underground PHS project in Illinois under the name of "Elmhurst Quarry Pumped Storage Project" (EQPS) makes use of an abandoned mine and quarry [18]. In addition to that, Riverbank Wiscasset Energy Center is also working on a 1000-MW underground PHS with a depth of 2200 ft underground located in Wiscasset, Maine [19].

Delft University has reported on studying a new innovation, which is compressed air combined with PHS technology [20]. A pressurized water container is used in this system to substitute for the upper reservoir; energy is being stored in compressed air instead of water at high elevation. The air gets pressurized when water is being pumped to the pressure container [20]. Thus, compared with PHS, this technology does not depend on specific geographic locations as much as PHS [21]. Another technology called undersea PHS has been proposed by Subhydro AS, which is a Norwegian company [22]. The usefulness of this concept is seen in the case of offshore wind farms. It basically uses the pressure of water at the ocean bottom [23]. The obtained results of this study have shown that the system can generate economic profits at relatively low depths of around 200 m. This technology has many advantages and potentials, although it still needs more research and investigation when it comes to the construction of the sphere. It also requires specific materials such as high-strength concrete. With all of these challenges on the way, undersea storage technology is considered a great innovation, which can be implemented [20].

In 1999, the first sea water PHS was developed in Japan under the name of "Yanbaru" [23]. The proposition came from Morishige at Mitsubishi Heavy Industries with the idea of large structures mounted at the sea bottom [24]. Other researchers suggested PSH systems, which can be mounted at the seafloor and coupled with offshore wind farms [25–27]. Unfortunately, these projects could not be implemented due to technical difficulties. On the other hand, Massachusetts Institute of Technology investigated another system named ocean renewable energy storage (ORES) [23]. Energy is stored in spheres located underwater. Two scientists from Germany (Schmidt-Bocking and Luther) also worked on a similar system. They suggested a storage technology with a capacity of 58,000 MWh, a sphere of 280 m in diameter, and a depth of 2 km under water [28].

There are many solutions that have been suggested by researchers to enhance PHS technology. Some focus on improvements related to elevation difference or to water pressure, whereas others pay more attention to the discharge rate of water. In the Netherlands, two brothers used a huge confine water area as PHS [21]. This system is also called "pump accumulation station" in which the fluid is being pumped to a certain elevation when there is excess generation and then released through turbines when energy is needed [21]. This system has been implemented by Kibrit in the Netherlands [29]. Another system called Energy Island has been developed by Boer and is still under research. This storage system operates by storing energy while lowering the water unlike the previously mentioned systems which elevate it [30]. A similar system named "The Eleventh Province" is being developed in Belgium [31].

To solve the elevation difference issue, Heindl came up with an original suggestion named Hydraulic Rock [32]. The concept is pressurizing water using a big piston unlike the idea of pumping water up or down depending on the geographic conditions. The implementation of this system is based on drilling a large circle to create a rock piston, which is isolated from the surrounding rocks [32]. The surface water and the underground area must be connected using a tunnel. This system proposes an ES capacity of 1600 GWh with the piston having a radius of around 500 m. Thus, this technology is considered as bulk storage as it can store large amount of energy. The large capacity of the system is an advantage; however, it is challenging to construct the assembly between the piston area and the structure [32].

A study by Hanley inspected the possibility of constructing a similar system suggested by Gravity Power, LLC [33]. Gravity energy storage (GES) technology is considered as one of the most interesting storage concepts because it relies on the same concept of PHS. The attractiveness of this system comes from the fact that it overcomes the constraints of site availability. The system operates using a piston, which is situated inside a large shaft that is filled with a fluid. Energy is generated by making water flow through the return pipe during the downward movement of the piston. The flowing fluid is used to spin a turbine which by itself is used to drive a generator. Mechanical energy is converted to kinetic flow energy when the system is in the storage mode, causing the piston to move upward. This operation is performed using a pump and a motor in the periods of low energy demand. The feasibility study of this system has been studied by Oldenmenger (2013) with a concept being investigated in a tall building [34]. GES has gained attention because it proved to be an environmentally friendly technology unlike other storage technologies such as batteries. Many studies have been investigating GES from different sides such as design and sizing [35]; other studies also reported on economic and risk analysis [34,36,37], whereas some researchers also focused on structural stability [38]. Yet, this system is still under research and needs more analysis.

ENERGY STORAGE TECHNOLOGIES

Currently, the increasing penetration of RE technologies into the grid is changing its operation modes. The electricity system faces many risks due to the variability of RE production. The main solution to this issue is the incorporation of ES systems to absorb the intermittency of energy from renewable sources. ES are able to flexibly store and discharge energy according to energy needs. Thus, they represent a key for mitigating the different issues that might occur to the electric system due to the high penetration of renewable systems. Besides, ES is very important when it comes to the economic and physical optimization of the electric system. In this chapter, different ES technologies along with their status will be discussed in detail.

Mechanical and hydraulic ES systems basically store energy in the forms of elevation, compression, or rotation. PHS is a well-established technology; however, it faces the challenge of finding the appropriate geographic location. CAES, on the other hand, was successfully implemented in Europe although it is not as widely used in the United States. This system can be implemented by exploiting empty natural gas reservoirs. Otherwise, energy might also be chemically stored as hydrogen in depleted gas reservoirs. Flywheels are used to store energy of rotation; the limitation of this technology is the need of advanced technical designs and materials, which must be used to optimize the cost and the storage volume. Because of inefficiencies, energy losses might occur up to 50% in mechanical and hydraulic storage systems.

Energy can also be stored through reversible chemical reactions. In theory, researchers are interested in storing heat at low temperatures in a chemical form; however, this idea has not yet been practically established. A similar example of this concept consists of storing hydrogen in metal hydrides; this is still under experimental testing phase. Although electrochemical ES systems have very good efficiencies, they are still very expensive. Thus, researchers are focusing on enhancing the performance and characteristics of batteries. One of the aspects that need to be improved is the ratio of the weight to storage capacity. Sodium-sulfur, lithium-sulfide, and lead-acid batteries, in addition to other types of batteries, are being the center of attention. Another type of electrochemical ES is called the redox flow cell. This type of batteries is charged and discharged through chemical reactions (oxidation and reduction) that take place at the level of two separate tanks. To facilitate the competitiveness of this system with the already existing ones, it is necessary to reduce its price by at least half of its current cost.

There are many systems and forms for thermal energy storage (TES); they may include bricks and ingots, designed containers, lakes and underground aquifers and soils. In Europe, there are some brick systems, in which energy is stored in the form of heat. On the other hand, thermal energy can also be stored as melted salts or paraffin. This is called latent storage, and it can enable reducing the volume of storage down to 100 times. This technology still faces technical issues that need to be solved. Last but not least, superconducting magnetic systems can be used to store electricity; however, such systems are very expensive.

With all the abovementioned technologies, there is room for many research topics to work on and develop more ES systems. The potential benefits of ES imply the huge need for research to achieve more efficient systems.

Among the R&D areas in the ES field there are
- Advanced ES systems
- Finding alternative ES
- Cost reduction of ES
- Design and optimization of new innovative systems
- Materials for ES
- Integration of ES into the grid

In the energy field, storage systems are considered one of the most important aspects. For example, storing oil is a way of storing the energy embedded in it. Storing oil is also very important to ensure the daily functioning of most of our activities because it provides a reliable production of energy through gasoline, petrochemicals, and fuel oil. For electricity, utilities use pumped storage to store energy. This storage scheme basically uses electric motors to pump water up to an elevation when excess energy is generated. At peak demand periods, water is released from the upper reservoir to a lower reservoir passing by a hydroturbine that drives a generator. Electricity can also be stored in batteries; an example of the use of batteries is in automobile applications for starting combustion engines.

ES is not only limited to electricity, as heat can also be stored. This kind of storage thermodynamically catches transferred heat before it is used. An example of heat storage could be hot water for residential and industrial applications. However, it can only be stored for short amount of time (less than 1 day). Developing advanced ES technologies has a lot of benefits especially when they are coupled with solar and wind systems. They are also crucial in developing efficient and environmentally friendly electric-powered vehicles. Therefore, there are a number of ES technologies that are under development. In this section, ES technologies will be discussed.

Pumped Hydro Storage

PHS represents the most reliable large-scale ES worldwide, which has been used since the 1920s [39]. It constitutes more than 99% of the globally installed large-scale ES [40]. Basically, this technology operates using the change in elevation for storing excess energy as a potential energy of water. This storage system has two reservoirs located at different elevation. At off-peak periods, when electricity is at its lower costs, PHS system is called upon to use the surplus energy to pump the fluid to the higher reservoir. When energy demand is high, water is released from the higher reservoir to the lower one. This water is used to spin a turbine connected to a generator, which produces electric energy. The amount of stored energy is determined by the volume of the stored water and the elevation difference between the upper and lower reservoirs.

Compared with any other ES technique, PHS has a lot of advantages. It is ready for use whenever electricity is needed within short periods of time; it is a very efficient and mature technology, and it is considered a cost-effective system. Depending on the application scale, PHS might reach efficiencies between 65% and 85% [41]. In addition, PHS has a very long lifetime of about 60 years [42]; it has low operation and maintenance costs and does not face the challenge of cycling degradation. Although PHS is a number of advantages, it also suffers from some constraints such as the need for an appropriate site, the high initial investment cost, and the long construction period it requires. This technology might face some environmental concerns. As for now, research has been focusing on finding innovative ways, which are based on the same concept as PHS but are able to overcome all the abovementioned challenges. Examples of such systems include underground PHS, seawater PHS, pump accumulation stations, subsea PHS.

The power output of PHS follows the basic fluid power equation which is expressed as:

$$P_O = QH\rho_W g\eta \qquad (1.1)$$

where P_O is the generated power; Q is the flow rate; ρ_W is the water density; H is the hydraulic head height; g is the gravitational acceleration; and η is the efficiency.

Compressed Air Energy Storage

This technology was developed and commercialized since the late 1970s [43]. It represents an alternative to PHS. CAES is developed for both small- and large-scale storage as well as commercial applications. Basically, this technology operates on storing energy as a form of compressed air. Excess electricity is stored at low peak demand periods using compressors to inject air into a storage vessel. When energy demand is high, the compressed air is heated and used by a steam turbine, which drives a generator to generate electricity. The source of heat used is usually either fossil fuels or recovered heat from the compression process. A recuperation unit can be used to recycle the exhaust waste heat.

PHS and CAES share a lot of similar advantages. Such characteristics include an interesting capacity, which can go up to 50 h, a high storage capacity of about 300 MW, and a rapid start-up. The fast response time is about 9–12 min. The system has also a high efficiency ranging from 60% up to 80% [44]. In addition, CAES technology has a long lifetime that can go from 20 years up to 40 years. CAES does not only share

the advantages of PHS but also the same challenges. They are both highly dependent on the geographic location, and both have low energy densities of around 122 kWh/m³ [45]. Thus, the most important limitation facing this technology is finding the appropriate implementation site. The most economically feasible locations are depleted gas fields or power plants closer to aquifers, rock mines, and salt caverns. On the other hand, compared with pumped hydro, CAES has a low capital cost that ranges between $400 and $800/kW. Moreover, the environmental impact of this technology is lower than that of PHS because it is usually underground [45]. However, CAES has some environmental concerns as it operates using natural gas.

CAES technology is very beneficial to the power system especially in the integration of renewable energies. It is able to smooth fluctuated wind power and can balance the issues due to the weather dependency and unpredictability of wind power. Thus, it represents a reliable option to the power utility operators. The total value of RE production is enhanced by the combination of compressed air storage with wind plants. Although this system has not been used widely, it can also be deployed both for large- and small-scale applications. Because CAES ensures power flexibility and grid stability, it will become more attractive in the future ES market. Fig 1.1 illustrates a sketch of diabatic CAES. Advanced adiabatic compressed air storage system is a new emerging CAES technology, which does not involve the burning of fossil fuels to expand the compressed air. Enhancing this technology for better efficiencies is still under research. Example of improvements includes small-scale CAES that are made of small vessels and CAES with humidification.

Flywheel Energy Storage

Although it is being integrated in the energy market at slow pace, flywheel ES is considered as a very interesting technology. It is one of the first mechanical storage methods. Flywheel ES uses kinetic energy as a form of storage. This technology has many advantages such as its high efficiencies ranging from 90% to 95%, its long-life cycles and long lifetime (15–20 years) [45]. Issues associated with this storage technology include the high capital cost of $1000–$5000/kWh and the high self-discharge rates that can go from 50% up to 100% [46]. Therefore, flywheel storage can be used especially for large storage capacities. In addition, because of its high self-discharge rates, this technology can be seen as effective only when storing energy for short periods of time. It is also used to regulate current fluctuations in power output from RE sources.

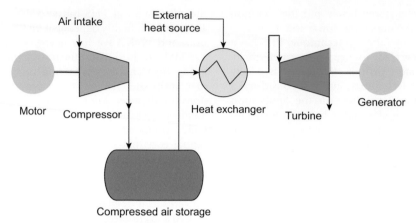

FIG. 1.1 Diabatic compressed air energy storage.

Flywheel stores energy in the form of rotational kinetic energy. It could be expressed as Eq. (1.2):

$$E_K = \frac{1}{2}I_m\omega_f^2 \text{ with } I = \frac{1}{2}m_f r_f^2 \tag{1.2}$$

where E_k is kinetic energy; I_m is the moment of inertia; m_f is the flywheel mass; ω_f is the flywheel speed; and r is the radius of the flywheel. Increasing the speed of the flywheel and its movement of inertia increases the energy production.

Capacitors and Supercapacitors

Energy is stored in capacitors by the accumulation of negative and positive charges. The system characteristics include a lifetime of about 5 years, an efficiency ranging from 60% to 70% [45], and a fast charging process. Capacitors that are used in power system applications for their ability provide a number of services such as power factor correction, VAR support, harmonic protection, and voltage support.

Supercapacitors are known as electrochemical capacitors, ultracapacitors, or electric double-layer capacitors. In 1950, the first supercapacitors have been created using activated charcoal plates. More efficient plates made of improved materials such graphene, carbon nanotubes, and barium titanate have been developed. A number of similarities exist between capacitors and supercapacitors. They both have a lifetime of about 8–10 years, as well as an interesting efficiency and power density. Supercapacitors can be recharged rapidly and are characterized by a good discharge performance and a high self-discharge [47]. These technologies are quite expensive compared with other ES systems.

The difference between capacitors and batteries resides in their ability to store large amount of energy compared with batteries. However, the storage period of this energy is short as opposed to their counterpart, which can store and discharge energy for a longer duration. Supercapacitors offer a number of services to the electric grid such as the improvement of power quality. Their ability to absorb high-frequency power deviations enables them to provide this service to RE plants [47]. These ES systems are more suitable for applications requiring high bursts of power.

The energy stored in capacitor is dependent on the voltage across the electrodes and the system capacitance (see Eq. 1.3). By increasing the surface area of the electrode, or by reducing the distance between them, the capacity of the system could be increased.

$$\begin{cases} W_C = \frac{1}{2}CV'^2 = \frac{1}{2}Q'V' \\ Q' = CV' \\ C = \frac{\varepsilon A_e}{d_e} \end{cases} \tag{1.3}$$

where W_C is the electrostatic energy; V' is the voltage; Q' is the charge; C is the capacitance; ε is the dielectric permittivity; A_e is the area of the electrode; and d_e is the distance between the electrodes.

Batteries

The working principle of batteries relies on the conversion of chemical energy using oxidation and reduction reactions to electric energy. There exist different types of batteries, and each of these types has their own characteristics and properties. The choice of a specific type of battery over another one highly depends on the application. Compared with other energy technologies, batteries are more expensive per energy unit. Moreover, there are some environmental concerns associated with the use of batteries. On the other hand, some of the technology advantages include a quick response time, space saving, and noiselessness. All of these properties make batteries a reasonable choice when it comes to smoothing fluctuations resulting from the integration of renewable sources into the electric grid. Thanks to the technologic progress, batteries are still subject to further development. In the following sections, different types of batteries will be described.

Lead-acid batteries

For more than 100 years, lead-acid batteries have been used for different applications such as residential and commercial applications [44]. Compared with some of the other batteries, lead-acid batteries still have some limitations, which make them not being widely used for commercial applications [41]. Because of their low cost, this type of batteries is used in diverse applications such as for uninterruptible power supplies, automotive starting, and lighting [6].

There exist various classifications of lead-acid batteries such as the valve-regulated battery, the flooded battery, and the sealed maintenance-free battery [39]. Research and development (R&D) in the field of ES focuses on enhancing the performance of lead-acid batteries to improve the storage lifecycle and discharge time. This is done through the use of innovating materials. Other research studies focus on the use of this type of batteries in the field of RE and automotive technologies. As for now, progress has been shown in the development of advanced lead-acid batteries such as "Ecoult UltraBattery smart systems."

The positive and negative electrodes of lead acid battery consist of Pb and PbO_2, respectively. The electrolyte solution is made up of H_2SO_4. The chemical reactions that occur at the anode and cathode are as follows:

At the anode:

$$Pb + SO_4^{2-} \leftrightarrow PbSO_4 + 2e^- \tag{1.4}$$

At the cathode:

$$PbO_2 + SO_4^{2-} + 4H^+ + 2e^- \leftrightarrow PbSO_4 + 2H_2O \tag{1.5}$$

The electrochemical reaction that takes place in lead-acid battery is expressed in Eq. (1.4):

$$Pb + PbO_2 + 4H^+ + 2SO_4^{-2} \leftrightarrow 2PbSO_4 + 2H_2O \tag{1.6}$$

$PbSO_4$ is produced during the discharging process with a release of water. The chemical process that occurs during the storage and discharging states of lead-acid batteries is shown in Fig. 1.2.

Lead-acid batteries offer a number of advantages. They are inexpensive (approximately $300−600/kWh); they provide high voltage (2 V), good reliability, and a high electrochemical effectivity (70%−90%). However, they have a short life cycle, which goes from several hundreds to thousands of cycles, and a low energy density. As this battery provides a number of benefits, it is now the most commonly used type of battery.

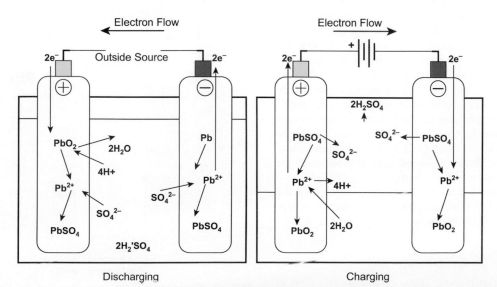

FIG. 1.2 Charging and discharging of lead-acid battery.

Lithium-ion batteries

This type of batteries is generally used in electric grid applications, electronics, and transportation. The cathode electrodes are made from graphite, and the negative ones are "lithiated" metal oxide; whereas the electrolyte is composed of a lithium salt such as lithium perchlorate ($LiClO_4$) or lithium hexafluorophosphate ($LiPF_6$), etc. [48]. Lithium-ion batteries have a high energy to weight ratio, a low self-discharge rate, and a life cycle of about 10,000. They have a very interesting efficiency of nearly 100% [48]. Li-ion batteries are very competitive because of their attractive weight and size compared with lead-acid batteries. They have a light weight, which represents one-fourth of the weight of lead-acid batteries [42].

R&D in the field focuses on the utilization of nanoscale materials to improve the performance of Li-ion batteries. In addition, other researchers investigate the development of new electrode materials and advanced electrolyte solutions to enhance the specific energy of batteries. As an example, advanced Li-ion batteries made from nanowires are able to produce 10 times more electric energy than that of conventional ones [42]. This type of batteries takes a share of more than half of the batteries used for portable devices; however, it is still facing some constraints such as its high cost. Therefore, the system manufacturing costs should be lowered to enable the spread of the technology in larger markets [45]. It is also important to mention the challenges facing Li-ion batteries, which include the heat requirement causing high self-discharge of the battery and corrosion resulting from the use of sodium.

The positive electrode of lithium batteries is made of a "lithiated" metal oxide such as lithium cobalt oxide ($LiCoO_2$), lithium nickel dioxide powder ($LiNiO_2$) or $LiMnO_2$, and lithium cobalt (III). Its negative electrode consists of graphite. Lithium salt forms the electrolyte solution. Examples of lithium salt are lithium perchlorate ($LiClO_4$) or lithium hexafluorophosphate ($LiPF_6$) dissolved in an "organic carbonate."

During the charging of the battery, lithium cations are transferred to the anode. This is later combined with charging electrons (e−) to form lithium atom. Executing the chemical process in reverse results in the discharging process of the battery.

The chemical reactions that take place during the charging and discharging processes are shown in Eqs. (1.7) and (1.8), respectively.

$$C + xLi^+ + xe^- \leftrightarrow Li_xC \tag{1.7}$$

$$LiMO_2 \leftrightarrow Li_{1-x}MO_2 + xLi^+ + xe^- \tag{1.8}$$

Sodium sulfur batteries

Sodium sulfur is another interesting type of batteries, which has emerged lately. The development of these technologies has experienced major growth from 1998 to 2008. Some of the characteristics of this battery include a high energy efficiency ranging from 75% to 90%, a cycle lifetime of about 2500, and a good power density [49]. This technology has an interesting Levelized cost of energy (LCOE) compared with other batteries, which makes it a good candidate for energy management. Some of the drawbacks of this technology include high-rated power cost of about $2000/kW and a high temperature requirement [45].

While passing through the electrolyte, positive sodium ions are converted into sodium polysulphides by combining with sulfur. Eq. (1.9) represents the chemical reactions that take place in the battery.

$$2Na + 4S \leftrightarrow Na_2S_4 \tag{1.9}$$

Na^+ ions pass through the electrolyte during the discharging process; while the electrons move through the battery external circuit. Fig. 1.3 illustrates the chemical process during the charging/discharging of sodium sulfur battery.

NaS batteries have become a commercialized product. These types of batteries have many advantages. Their abilities to deliver power in a continuous discharge and in shorter pulses remain the greatest one. NaS batteries are formed from sodium that has a low atomic weight and a high reduction potential,

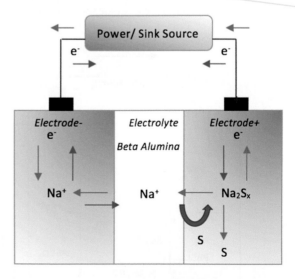

FIG. 1.3 Charging and discharging of sodium-sulfur battery.

which allows the battery to have a high specific energy of about 100–200 Wh/kg. The cathode is made of molten sulfur, which is cheap and available in nature. The anode is made of the sodium which is nontoxic and nonexpensive. The separator is a β-alumina membrane ceramic electrolyte.

During the discharge process, the electrolyte allows sodium ions to flow from the negative electrode to the positive one. Once the sodium is combined with sulfur, sodium polysulfide is produced. The process is reversed during the charging process. Sodium ions are converted to sodium at the cathode, while sodium polysulfide decomposes.

The discharge reaction occurs in two steps:

$$2S + 2Na \rightarrow Na_2S_5 \tag{1.10}$$

$$3Na_2S_5 + 4Na \rightarrow 5Na_2S_3 \tag{1.11}$$

NaS batteries are made up of nonexpensive materials that are abundant. These batteries could be a practical solution for smoothing the intermittent power produced by wind plants.

Na-metal chloride battery. Similar to NaS batteries, the separator of this battery is also made of ceramic β-alumina. It has also a second electrolyte, which is molten sodium chloroaluminate ($NaAlCl_4$) to create an appropriate ionic contact between the cathode and the electrolyte. The negative electrode is composed of molten sodium, whereas the positive one consists of metal chloride. This is a combination of sodium chloride and metal powder.

The possible occurrence of both an overcharge and an overdischarge reaction makes this battery more advantageous than sodium sulfur cell. An overcharge occurs when molten sodium chloroaluminate electrolyte reacts with metal; while an overdischarge occurs when the aforementioned electrolyte reacts with sodium.

Reactions occurring during the discharge process are as follows:

$$2Na + NiCl_2 \rightarrow 2NaCl + Ni \tag{1.12}$$

$$2Na + FeCl_2 \rightarrow 2NaCl + Fe \tag{1.13}$$

The overcharge reaction and the overdischarge reaction for sodium nickel chloride cell are given below:

Overcharge reaction for sodium nickel chloride cell:

$$2NaAlCl_4 + Ni \rightarrow 2Na + 2AlCl_3 + NiCl_2 \tag{1.14}$$

Cell overdischarge reaction is

$$3Na + NaAlCl_4 \rightarrow Al + 4NaCl \tag{1.15}$$

Metal-air batteries

This battery type is still an emerging technology. It is considered a type of fuel cell that uses air as the oxidizing agent and metal as the fuel. This battery option has many advantages over other storage systems. It has a higher energy density and a longer lifetime and is considered an economic storage solution. However, it also has some downsides that need to be resolved in order for this technology to play a major role in the energy market. These major drawbacks are related to the system poor recharging capacity and low efficiency, which is less than 50% [45].

The anode of metal air batteries is made of metals such as zinc or aluminum, which have high energy densities. When oxidized, these later can release electrons. The battery cathode consists of a metal mesh or a porous carbon structure enclosed by good catalysts [45]. The electrolytes that could be in liquid state or solid polymer form should usually have a good hydroxide-ion conductivity. An example of such electrolyte is KOH.

Zinc-air battery reactions are shown in Eqs. (1.16)–(1.18):

At the anode:

$$Zn + 4OH^- \rightarrow Zn(OH)_4^{2-} + 2e^- \tag{1.16}$$

At the cathode:

$$O_2 + 2H_2O + 4e^- \rightarrow 4OH^- \tag{1.17}$$

Overall reaction:

$$2Zn + O_2 \rightarrow 2ZnO \tag{1.18}$$

Nickel-based batteries (NiCd, NiMH, and NiZn)

Battery systems that use nickel-electrode are nickel-cadmium (NiCd), nickel-metal hydride (NiMH), nickel-iron (NiFe), nickel-hydrogen (NiH_2), nickel-zinc (NiZn). The most used one in utility industries is nickel-cadmium (NiCd). However, NiCd and NiMH are commonly used than the other types. The cathode electrode consists of nickel-hydroxide, whereas the electrolyte of nickel-based batteries is an aqueous solution of potassium hydroxide and lithium hydroxide [46]. The anode electrode is made up of cadmium hydroxide, zinc hydroxide, and metal alloy for NiCd, NiMH, NiZn batteries, respectively. NiCd battery storage has a cycle lifetime of about 1500–3000 cycles [46]. Yet, it has a longer lifetime than lead-acid batteries, whereas NiMH and NiZn have smaller or almost the same calendar life time. The downsides of this technology are high self-discharge rates, which is approximately 10% per month, and high cost. Nickel-based batteries are 10 times more expensive and have lower efficiency range than lead-acid batteries [46]. In addition, the toxic "cadmium" material has a negative impact on the environment [45].

The positive electrode of nickel-based batteries consists of oxyhydroxide, whereas the negative one is made up of any metal Fe/Cd/Zn. The electrolyte solution uses potassium hydroxide. The overall electrochemical reaction of nickel-based batteries is shown in Eq. (1.19).

$$X = Fe/Cd/ZnNi$$
$$X + 2NiO(OH) + 2H_2O \leftrightarrow 2Ni(OH)_2 + X(OH)_2 \quad (1.19)$$

$Ni(OH)^2$ and $Fe/Cd/Zn(OH)^2$ are produced during the discharging process. The chemical process that occurs during the storage and discharging states of nickel-based batteries is shown in Fig. 1.4.

Nickel-cadmium battery. As any other batteries, NiCd batteries have some advantages as well as some disadvantages. They are characterized by high energy density, short response time, and high efficiency. However, they have a negative impact on the environment and require a certain safety control and cycling. NiCd batteries are more expensive than lead-acid batteries ($600/kW). The positive electrode of NiCd batteries is nickel oxyhydroxide, whereas the negative one consists of metallic cadmium. A nylon divider separates the two electrodes. The combination of nickel oxyhydroxide with water results in hydroxide and a hydroxide ion during the discharge phase. During the storage phase, the production of oxygen occurs at the positive electrode, whereas hydrogen is produced at the negative one. The electrochemical reaction that takes place is expressed in Eq. (1.20):

$$2NiOOH + Cd + 2H_2O \rightarrow 2Ni(OH)_2 + Cd(OH)_2 \quad (1.20)$$

NiMH battery. This type of batteries has more energy density than NiCd batteries (25% higher). It has a power density of around 200 W/kg, a cycle life of more than 1000 cycles, and is not very harmful to the environment. There are some differences between this battery and the NiCd one; the nickel metal hydride cell is characterized by a high self-discharge and cannot tolerate overcharge.

The positive electrode consists of NiOOH, whereas the negative one is hydrogen absorption alloys. Alloys comprise two metals; the hydrogen is absorbed exothermically by the first metal and endothermically by the second one. The separator used in the NiMH cell is the hydrophilic polypropylene. The reaction that occurs during the discharge is

$$NiOOH + MH \rightarrow Ni(OH)_2 + M \quad (1.21)$$

This kind of batteries is used in many sectors and applications with a long cycle life; such as electric razors, mobiles phones, toothbrushes, hybrid electric vehicle batteries, cameras, and others.

NiH₂ battery. NiH_2 is a hybrid technology that combines both battery and fuel cell systems. It is an alkaline battery used mainly in satellites. It has a long lifetime, a standard voltage of 1.32 V, and a higher specific energy (50 Wh/kg). The development of this battery is mainly to replace the NiCd batteries used in space applications.

This battery makes use of two types of separators, which include the asbestos paper and the porous ceramic paper. Potassium hydroxide electrolyte is

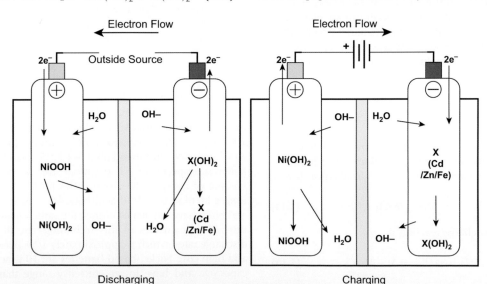

FIG. 1.4 Charging and discharging of nickel-based battery.

absorbed by the separators. The negative electrode is made up of platinum black catalyst, and the positive one is made up of nickel oxide.

The reaction occurring during the discharge state of the battery is

$$2NiOOH + H_2 \rightarrow 2Ni(OH)_2 \qquad (1.22)$$

During the charging process, the released hydrogen gas is stored in a cell pressure vessel. The pressure of hydrogen increases to 4 MPa during the charging process; whereas it decreases to 0.2 MPa during the discharged process.

Flow batteries

Flow batteries have gained popularity for their ability to operate at nearby ambient temperatures. They are less expensive compared with other ES systems. There are two types of flow batteries, which include redox and hybrid flow batteries. Vanadium redox batteries (VRBs) are the most common ones for the redox (oxidation/reduction) type; whereas zinc-bromine batteries are most commonly used for the hybrid type. The key difference between flow and conventional batteries resides in the fact that conventional batteries store energy as the electrode material, while energy is stored as the electrolytes in flow batteries. Because of this special property, flow batteries can be quickly recharged as there is a possibility for replacing the electrolyte.

Flow batteries have similar advantages to those of fuel cells and electrochemical accumulator cells. Such advantages include a separable liquid tanks and almost unlimited durability over conventional batteries. However, the challenges facing flow batteries include the high cost and the low energy density of the system. VRBs have some benefits, which include high energy efficiency that can reach up to 85%; long lifetime of 1250 cycles and up to 12 years; and a short response time. Although the zinc-bromine flow batteries have relatively lower efficiencies of about 75% and lower costs, the components that made up of both these types of flow batteries are the same [50].

Energy is stored in vanadium redox flow battery (VRFB) by exchanging electron between the ionic vanadium materials. The storage system charges energy by accepting an electron and hence converting V^{3+} to V^{2+} at the anode. Energy is discharged by releasing an electron and converting V^{2+} back to V^{3+}. At the cathode, the same process occurs between V^{5+} and V^{4+}. Fig. 1.5 illustrates the charging and discharging processes of a VRB. The reactions can be simply expressed as the following:

$$\text{At the positive electrode: } V^{4+} \leftrightarrow V^{5+} + e^- \qquad (1.23)$$

$$\text{At the negative electrode: } V^{3+} + e^- \leftrightarrow V^{2+} \qquad (1.24)$$

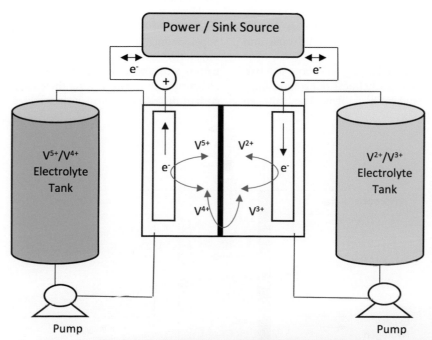

FIG. 1.5 Charging and discharging of vanadium redox flow battery [19].

VRBs can be used in many applications because they have large capacities. They may be used to balance the production of highly unstable power sources as RE plants. VRBs have an efficiency of 85% and a lifecycle of 5—10 years [48]. They have also some disadvantages such as low charge efficiency, small energy density, and high price and have a negative impact on the environment as they produce toxic leftovers.

Br₂—Zn battery. The zinc-bromine battery consists of the bipolar electrodes made from carbon-plastic composite material. The separator is a microporous polyolefin membrane, which permits ions to pass through it. The zinc is placed on the anode electrode, and the bromine is released from the cathode electrode during the charging process. In the discharge process, Zn and Br combine to produce bromide. The reaction occurring during the discharge process is expressed as

$$Zn + Br_2 \rightarrow ZnBr_2 \qquad (1.25)$$

Hydrogen Fuel Cells

Hydrogen has seen an increasing popularity and will play a vital role in the future. It is not only used to generate electricity; it has also become an attractive option to fuel hybrid vehicles. During the charging process, hydrogen fuel cells use excess energy to generate hydrogen. Hydrogen feeds the fuel cell to create electricity during peak demand (discharging process). Hydrogen is produced by a process known as electrolysis. The excess electricity is used to split water into hydrogen and oxygen.

The structure of a fuel cell is very analogous to a battery; it consists of an electrochemical cell comprising a cathode, an anode, and an electrolyte. The hydrogen ion transfer occurs within the electrolyte. The electrolyte and the reactant of the fuel cell can vary as with batteries. Unlike a battery, the fuel cell is not a closed system. In other words, the fuel cell keeps operating as long as the reactants are supplied, whereas the amount of stored chemical energy is limited within a battery. As flow batteries, fuel cells have the same advantage of separating the power rating and the energy capacity of the device. One disadvantage of this system is the low round-trip efficiency of the system, which is about 59%. Fuel cell is then considered less efficient than many other ES systems. Furthermore, this technology requires some time to start operating, which makes it inappropriate for applications that require a quick response time. Fuel cell's chemical reaction is shown in Eq. (1.26).

$$2H_2 + O_2 \leftrightarrow 2H_2O + \text{Electricity} \qquad (1.26)$$

Hydrogen fuel cell chemistry process is illustrated in Fig. 1.6. The hydrogen fuel dissociates to form hydrogen and electrons. The hydrogen ions flow to the oxygen electrode passing through the electrolyte. At the same time, electrons power the load by passing through the battery's external circuit. The combination of hydrogen ions, electrons, and oxygen produces water. In the regenerative process, the cell is fed by oxygen and hydrogen resulted from the separation of water by a power electrolyzer. Therefore, water and electricity are produced by this process.

Superconducting Magnetic Energy Storage

Superconducting magnetic energy storage (SMES) is an innovative system that stores electricity from the grid within a magnetic field that is created by the flow of DC current in a coil. During the charging process, the current increases, whereas it decreases in the discharge

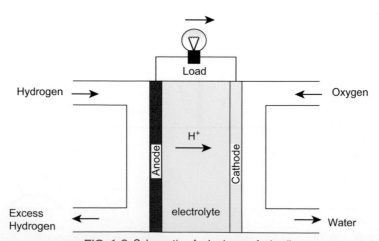

FIG. 1.6 Schematic of a hydrogen fuel cell.

process. In the standby application, the current circulates the coil. Therefore, magnetic energy can be stored indefinitely. SMES consists of a magnetic storage unit and a cryostat. This technology has the ability to store and discharge large amount of energy instantaneously. It is capable of releasing high amount of power within a fraction of a cycle. SEMS plays a crucial role in providing energy during momentary power outages and voltage sags. With the increasing penetration of RE such as solar and wind, injection of momentary bursts of power is necessary to maintain grid reliability.

The system has high efficiencies of approximately 95% and low loss of charge of about 0.1%. Because this technology is capable of responding in less than 100 ms, it is a great option for load leveling between the transmission network and RE sources. However, this technology is still expensive making it applicable for only short-term storage. The power units of SEMS that are currently used are of the order of 1−10 MW with a storage time on the order of seconds. SEMS units that can provide up to 100 MW are under development, storing energy on the order of minutes. Large MW units require extra safety measures to avoid exposing humans to harmful magnetic fields.

The energy stored by this technology is given in Eq. (1.27). This energy is a function of the current that flows through the coil and its self-inductance.

$$W_L = \frac{1}{2}L_c I_c^2 \qquad (1.27)$$

where W_L is the coil stored energy; L_c the self-inductance of coil; and I_c is the current.

Thermal Energy Storage

TES stores electricity or other waste heat sources in the form of thermal energy. These storage systems are classified into three categories, which include sensible heat, latent heat, and thermochemical heat. The method used by the first aforementioned category is the change in material temperature, while the phase change of a material is used by latent heat. The last category induces thermal changes in a material's chemical structure. The selection of the appropriate TES method depends on several factors, which include the system application, the storage system temperature range, and media. Thermochemical heat storage uses the ideally reversible phenomenon of sorption. A storage material (the sorbent) is capable of storing or returning heat by desorption or sorption of a fluid (sorbate). The sorbent, the sorbate, or their mixture can be stored in thermodynamically stable forms at room temperature, thus allowing long-term storage.

TES systems are also classified into high- and low-temperature systems depending on the operating temperature range. If this is below 200°C, the system is operating in low temperature. This TES storage category has been developed extensively and is used in cooling and heating applications of buildings. High-temperature TES category is mostly used in RE applications, thermal power systems, and waste heat recovery. Eq. (1.28) illustrates the thermal energy stored in sensible heat storage systems. These depend on a change of the temperature. The mass of the medium and the specific heat affect the storage capacity.

$$E_{th} = K_h(T'_2 - T'_1)V_s \qquad (1.28)$$

where E_{th} is the stored thermal energy; V_S is the volume of the system; T'_1 and T'_2 are the initial and final recharge temperature; and K_h is the specific heat.

Gravity Energy Storage

GES is a new storage technology that is under research and development. There are currency only small and laboratory prototypes of this system. Similar to PHS, this storage concept relies on gravity to store electricity. A number of constraints faced by PHS have been resolved by GES, such as the huge water requirement and the geologic limitations. Fig. 1.7 shows a schematic

FIG. 1.7 Schematic of gravity storage.

of this storage technology. The system components include a large shaft filled with water, a heavy piston, a return pipe, and a powerhouse. This later is linked to the return pipe; it consists of a motor-generator and a pump-turbine assembly. To store energy, the pump consumes excess electricity to push water to move in a counterclockwise direction; this results in an upward movement of the piston. During the discharging phase, due to the piston downward motion, water is forced to move in the reverse direction passing through the turbine, which drives a generator.

Comparison of Energy Storage Technologies
Each ES system has its own unique characteristics, which enable it to be used for a specific storage application. The identification of suitable ES is based on the system features, which include energy and power density, discharge time, power rating, lifetime, efficiency, cost, and others. Tables 1.1 and 1.2 present a comparison of the technical characteristics for some of the discussed ES systems [41,48].

NOVEL ALTERNATIVE DESIGNS OF PUMPED HYDROSTORAGE
This section discusses other storage concepts that rely on the same principle used by PHS. These technologies make use of hydraulic equipment, which include a reversible pump/turbine and a motor/generator to convert the gravitational potential energy into electric energy and vice versa. These systems are capable of complementing PHES by expanding their use in regions where the land is not appropriate for conventional pumped hydro or where public opposition can make PHES projects impractical, for example, regions with beautiful landscapes or agricultural lands. In addition, these new technologies may offer other applications to meet the need of the electric grid, which witnesses a high penetration of RE system requiring longer storage/discharge duration or higher energy demands compared with PHES. There are multiple PHES designs, among which the following designs are:

Underground PHES: A different configuration of PHES reservoirs has been proposed by researchers for a long time. This includes the use of underground caverns as lower reservoirs. However, such a system has never been constructed yet. In the United States, interest in developing underground pumped hydro has emerged again. A number of US developers have acquired preliminary authorization to investigate the practicality of building underground pumped hydrostations at their selected sites.

Variable-speed PHES: The variable-speed PHES technology is being actively introduced in European countries to integrate more variable renewable electricity in their power profiles. Fixed-speed pump turbines are present in most of the existing PHES stations. Although PHES, which makes use of fixed-speed pump/turbine, represents a cost-effective storage capacity, this system offers frequency regulation service only during its generating mode. The demand for frequency regulation has been increasing with the growing adoption of RE systems. With the use of variable-speed systems, PHES plants can regulate frequency during both generating and pumping modes. An example of PHES plant equipped with variable-speed systems is the Okawashi station in Japan.

Compressed air PHES: The upper reservoir of a traditional PHES can be replaced by a pressurized water container to obtain another innovative storage system. When water flows through the container, the air within the pressure vessel is pressurized. That is how this system stores energy in compressed air instead of storing it as potential energy in the elevated water. This novel concept could solve the geographic limitations faced by PHES systems and make them practical for almost any site (see Fig. 1.8).

Seawater PHES: A number of researchers are suggesting new systems similar to PHES with the growing worldwide interest in installing more PHES. Okinawa station is considered the world's first seawater PHES system that was put into operation by Japan in 1999. The station is composed of an above sea—constructed upper reservoir and a lower reservoir, which make use of open sea. One interesting approach to seawater PHES system, known as the energy islands, was studied by DNV KEMA company. The concept uses the open sea as the elevated reservoir, whereas the lower reservoir is obtained by excavating and constructing an undersea ring-dyke.

Undersea PHES: It is a novel idea that stores the electricity produced by offshore wind turbines using the water pressure at the bottom of the sea. Pressure vessels, in the shape of hollow tanks made of concrete, are placed on the seabed. Excess energy is stored by forcing water to flow out of the vessel, while it is generated by the filling of the vessel with seawater. Fig 1.9 shows the charging process of ORES.

Energy is stored in seabed hydrostorage by pumping water out of a pressure vessel. This is made of concrete and is fixed to the seabed at approximately 400 m depth and more. This energy is produced when water flows back into the vessel passing through a hydroturbine. 1 kWh of energy can be produced by a 1 cubic meter

TABLE 1.1
Energy Storage Technical Characteristics.

	PHES	CAES	Flywheel	Supercapacitor	SEMS	GES	Fuel Cell
Energy density (Wh/kg)	0.5–1.5	30–60	10–30	25–15	0.5–5	1.06	800–10,000
Power density (Wh/kg)	–	–	400–1500	500–5000	500–2000	3.13	500
Power rating	100–5000 MW	5–300 MW	0–250 kW	0–300 kW	100 kW–10 MW	40–150 MW	0–50 MW
Discharge time	1–24 h+	1–24 h+	ms–15 min	ms–60 min	m–8 s	34 s	s–24 h+
Storage duration	h–month	h–month	s–min	s–h	min–h	h–month	h–month
Lifetime	40–60	20–60	15	–	+20	30+	5–15
Cost ($/kW)	600–2000	400–800	250–350	100–300	200–300	1000	+10,000
Cost ($/kWh)	5–100	2–50	1000–5000	300–2000	1000–10,000	–	–
Cost ($/kWh-per cycle)	0.1–1.4	2–4	3–25	2–20	–	–	6000–20,000
Efficiency	65–87	50–89	85–95	90–95	95–98	75–80	20–35
Maturity	Matured	Developed	Commercial	Developed	Commercial	Concept	Developing

TABLE 1.2
Comparison of Batteries.

	Specific Energy (Wh/kg)	Specific Power (Wh/kg)	Cycle Life	Self-Discharge	Energy Efficiency
Lead-acid	30–50	75–300	500–1500	2%–5% per month	80
Nickel cadmium	75	150–300	2500	5%–20% per month	70
Lithium-ion	150–200	200–315	1000 –10,000	%1 per month	95
Sodium-sulfur	150–240	90–230	2500	–	Up to 90
Sodium-nickel chloride	125	130–160	2500+	–	90

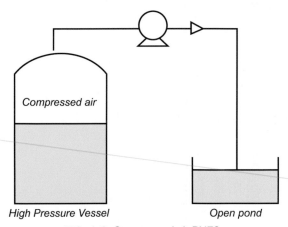

FIG. 1.8 Compressed air PHES.

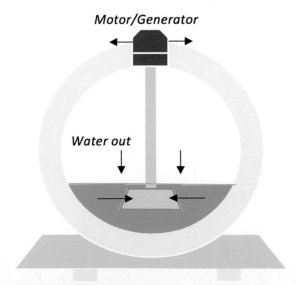

FIG. 1.9 Charging of ocean renewable energy storage.

system located at about 400 m depth. The operating pressure of the system is 4 MPa. Several researchers are investigating a number of aspects associated with this storage system. The Massachusetts Institute of Technology considers the use of concrete spheres with a diameter of approximately 25–30 m located at a depth of 400 m. The storage capacity of this system is around 6 MWh. The optimal and the most economic depth for installing such technology is 750 m. The cost of installation of this system is around $60/MWh [51]. The designed system has the potential to operate at larger depths with a wall thickness reaching 3 m. The concrete mass can counteract the buoyancy effect, helping the system sit on the seabed without the need for complex anchoring. However, the weight of the tank can be an obstruction as it is around 10,000 t.

SINTEF and Subhydro have set up projects in Norway, which bring advanced concrete technology to the fore. The aim is to reduce the weight and cost of the vessels to be able to fix them to the operating site. Cylindric tanks are used as pressure vessels, where the design volume varies according to the length of the cylinder. A number of units are grouped to form an interconnected system. Air at atmospheric pressure is provided to the tank by a ventilation shaft. This system may have other functions.

Underwater PHS is meant to be paired with offshore wind turbine; this hybrid system could provide day-ahead, power smoothing, and storage services. It could also offer other services and applications especially in regions that are close to the shore and have convenient depths.

Other innovative hydroelectric storage systems that will be discussed in this chapter include hydraulic piston ES, offshore lagoon PHS, energy membrane storage, novel PHS, and energy island storage. The first two technologies are similar in principle to hydraulic accumulators as they also use the techniques used by pumped hydrotechniques to raise a solid mass. The last two technologies make use of water as the working mass. The design of these systems is different from the traditional PHES with regard to some details.

Hydraulic piston ES: The use of a piston as a mean of storing energy by lifting a heavy mass has been proposed a number of times. The proposed systems have different configurations with regard to the piston, the reservoir, and the power house. Groundbreaking energy storage (GBES) system can be combined with other energy technologies. It can significantly increase the capacity of an existing PHES system. When integrated to a conventional hydroelectric station, GBES systems can increase its ES capacity without requiring another reservoir. They can be integrated into tidal barrage or lagoon stations; they can also be sited between offshore wind turbines. GBES systems can house a large public water supply reservoir or store excess seasonal water. It can also be used for flood relief. Water management and energy supply are two of the most crucial challenges of this century. Having a technology that deals with the aforementioned issues is significantly important in gaining support and reducing costs. During the past 10 years, a number of concepts sharing the same principle as GBES were implemented.

Initially, GBES system was started as an idea by Esco-Vale in 2007 for gigawatt-class storage at the seafloor, using pistons of approximately 500 m diameter installed within the offshore arrays of wind turbine. The land-based application of this technology includes imitations of PHES in flat areas and future markets that other electrical ES systems cannot be part of, not even most PHES plants. Additionally, one of the oldest and most developed firms is Gravity Power (GP). It makes use of a high-density concrete piston that has a relatively small diameter. The system uses fabricated or geologic pistons. This moves within a vertical deep shaft to store and discharge energy.

Heindl Energy (HE), on the other hand, proposed the use of geologic pistons. They suggested systems with the same scale as large pumped hydroplants with rated power that can reach gigawatts with hundreds of gigawatt-hours energy. HE plants included particularly large systems designed for large capacity seasonal storage.

PHS-subterranean reservoir: It has been mentioned before that there have been many previous proposals with very few being successful, for the development of pumped hydroprojects where the power house along with the lower reservoir is placed underground. The elevation head is often similar to that of a PHES plant, but there are no specific limitations. In South Africa, some mines extend over thousands of meters of depth. The major restrictions consist of the underground chamber's capacity and the mine workings stability due to the fact that water flows rapidly through the tunnels and the shafts.

Energy membrane storage: The energy membrane-underground PHS was developed by JolTech in Denmark. Energy is stored in this system by pumping the fluid into a cavity that is surrounded by the membrane layers. The energy is reproduced by releasing the water, which flows through a turbine. The suggested commercial design has a rated power of 30 MW and a storage capacity of 200 MWh. The cavity is concealed underneath soil at 25 m. EM-UPHS is intended to operate in sites where the topography is not appropriate for the conventional pumped hydro, but it can make use of the height difference between the source of water and the cavity. The storage capacity and the system operating pressure can be increased by 10% by the addition of 5 m. A pilot project, which has been operated over several years, has helped in the refinement of the concept. The proposed pilot plant is a small scale of the commercial system.

Shallow water lagoon ES: Because of the lack of freshwater resources, the implementation of PHES might not be successful even in regions with a convenient geology. Seawater may be used to solve the aforementioned issue. The main drawbacks of using seawater include the high corrosion rate, the low efficiency, and the high specific cost of the low-head hydrosystem. If ocean power is more used to produce energy, saline low-head hydroelectric technologies will have more chances to be developed.

An enclosure is built around a considerably large area of water in a region characterized by high tidal range to establish a basic tidal power generation lagoon. Before low tides, water is kept within the lagoon; it is then discharged through a turbine producing energy. During high tides, water is allowed to reenter the lagoon producing more energy.

It is logical to think about including a pump storage option in barrage and tidal lagoon systems. At high tides, pumping more water into the lagoon does not require much energy because the water is raised for a small elevation. If at ebb tide the same water is allowed to flow back, it will produce a lot more energy. For example, if at high tide, 10 MWh of input energy is needed to raise the lagoon level by 1 m, then 100 MWh of additional energy is expected to be retrieved when the water is excluded at ebb tide. This method can be reproduced at low tide by forcing water to flow out of the lagoon, decreasing its height by 1 m.

Energy island storage in deeper water: In deeper water, a similar approach can be applied. The technology would be ideally located near the shore but should not be greatly influenced by tidal enhancement. A large area is usually enclosed by a constructed barrier of about a few km^2 to 100 km^2 to develop an internal lagoon island with a water depth of 50 m. Water is pumped out of the lagoon to store energy; this process lowers the lagoon level to about 45 m. The energy is retrieved as water flows back to the lagoon passing through a turbine. Theoretically, the full discharge of the system is possible and would result in water rising to a level beyond the island. This would be impractical because a small quantity of energy would be retrieved in the final stages. Most proposed systems are supposed to operate within a smaller head range of around 40 m.

Advanced rail ES: The considerable and continuous growth of electric storage market due to the high integration rate of renewable resources has led to the development of innovative ES systems. Advanced Rail Energy Storage (ARES) company proposed an energy-efficient alternative method of utility-scale electric storage. The system makes use of a fleet of electric shuttle trains carrying heavy masses between two yards located at different elevations. It operates with low-friction automated steel rail. This novel system has been developed by ARES LLC firm in collaboration with major companies in the railroad and energy sectors.

ARES is an ES system established on a rail concept. It is considered similar to pumped storage hydroelectric system in the fact that it stores electricity by increasing the position height of masses. This stored energy is retrieved as the mass gets back to its initial position. During the period of low energy demand, excess electricity is used by ARES shuttle trains to power its axle-driven motors. This transports the masses uphill to the upper yard; against gravity force. During peak periods, the aforementioned process is reversed. The flow of masses is transported back to the lower storage yard by the use of generators. The process of converting the masses elevation potential energy to electricity is efficient. The system efficiency has been approximated by ARES LLC to 80%. To develop a highly reliable system, ARES firm integrates the most advanced power control systems with recent generator and motor traction-driven technologies. The system is characterized by a scalable energy and power, which enables it to participate in diverse applications.

To meet the grid energy needs, ARES stations can be developed over a large range of power rating and energy capacity. The system can be implemented in mountains and hilly regions without any major impact on the environment. The storage capacity of this technology ranges from 100 MWh to few GWh. ARES technology has a low capital cost (almost double cost) compared with its competitors such as pumped hydroelectric storage, while it can deliver the same amount of power and energy.

A pilot project was developed and tested in California, whereas a commercial system is being developed in Pahrump, Nevada. This system is considered the most environmentally friendly technology among other ES options. ARES does not use any fuel, and it is a zero-emission technology. In addition, the system does not make use of any dangerous or polluting substances and can be taken out of use without any permanent environmental impact.

ENERGY STORAGE STATUS WORLDWIDE

A number of ES projects with varying scales have been developed globally, as shown in Fig. 1.10. Currently, the highest number of operational ES projects belongs to battery ES (more than 350). PHES is ranked in the second place with approximately 300 projects. This is followed by TES systems [41].

PHS constitutes about 98% of the amount of energy stored as seen in Fig. 1.11. This is followed by thermal storage and flywheel, which represent approximately 1% each. The share of capacitors, hydrogen, and CAES is very small and less than 1% [41].

The operational battery projects are classified into different types of batteries, which have been shown in Fig. 1.12. The most commonly used type is Li-ion battery storage with more than 100 projects followed by Li-ion phosphate, and sodium-sulfur batteries. However, the share of Li-ion battery in terms of amount of energy stored is less than that of sodium-sulfur as illustrated in Fig. 1.13. Li-ion battery represents only 18% compared with 24% of NaS batteries. This later is mostly used for large-scale energy application while Li-ion battery is typically used for portable applications. Advanced lead-acid battery constitutes an interesting share of 18% compared with other batteries. This battery type has only 15 projects currently operational worldwide. This shows that advanced lead-acid batteries are used for large-scale applications.

Concerning TES systems, the highest number of operational thermal plants in the world, goes to ice thermal

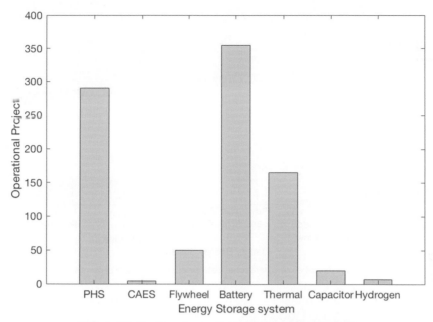

FIG. 1.10 Number of operational energy storage projects.

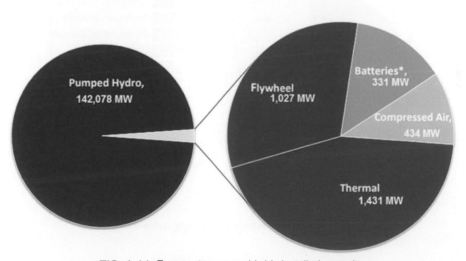

FIG. 1.11 Energy storage worldwide installed capacity.

storage followed by molten salt (see Fig. 1.14). Chilled water thermal storage is in the third place followed by heat storage. However, the quantity of energy stored of molten salt is higher than that of ice thermal. This represents only 4% compared with 77% of molten salt as illustrated in Fig. 1.15. The reason behind this high difference is because ice thermal storage is used for small-scale applications such as air conditioning; while TES with molten salt is used for large-scale application such as CSP solar power plants [41].

The most used type of CAES system is in-ground natural gas, as illustrated in Fig. 1.16. The two currently operational projects of this type include McIntosh and Huntorf CAES. As for the two remaining types of CAES, they both have only one project currently operational.

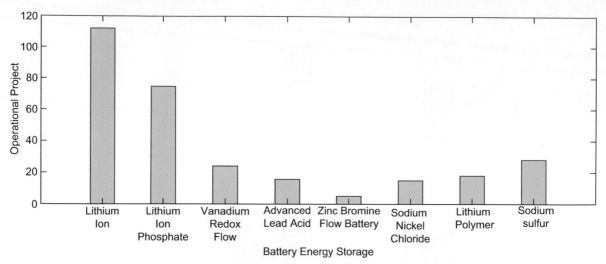

FIG. 1.12 Number of operational battery energy storage projects.

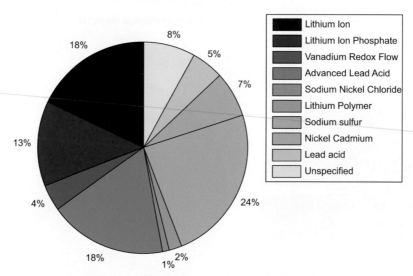

FIG. 1.13 Share of energy stored using batteries.

CONCLUSION

ES has been suggested as a viable solution to overcome the variability issues faced by RE systems. This chapter has presented an overview about the main types of ES methods in existence outlining their characteristics. The current ES projects worldwide have also been presented. Finally, a brief discussion about RE and ES projects in Morocco has been discussed.

NOMENCLATURE

A_e Area of the electrode
C Capacitance
d_e Distance between the electrodes
E_k Kinetic energy
E_{th} Stored thermal energy
g Gravitational acceleration
H Hydraulic head height
I_c Current

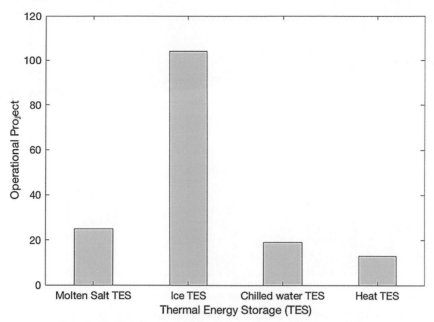

FIG. 1.14 Number of operational thermal energy storage projects.

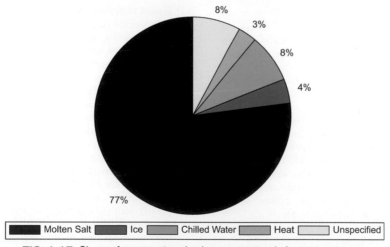

FIG. 1.15 Share of energy stored using compressed air energy storage.

I_m	Moment of inertia	T'_2	Final recharge temperature
K_h	Specific heat	V_S	System volume
L_c	Self-inductance of coil	V'	Voltage
m_f	Flywheel mass	W_C	Electrostatic energy
P_O	Generated power	W_L	Coil stored energy
Q	Flow rate	η	Efficiency
Q'	Charge	ε	Dielectric permittivity
r	Radius of the flywheel	ρ_W	Water density
T'_1	Initial recharge temperature	ω_f	Flywheel speed

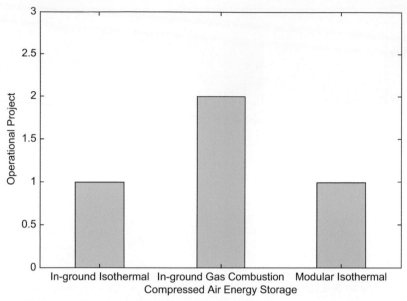

FIG. 1.16 Number of operational compressed air energy storage projects.

REFERENCES

[1] Mahlia T, Saktisahdan T, Jannifar A, Hasan M, Matseelar H. A review of available methods and development on energy storage; technology update. Renew Sustain Energy Rev 2014;33:532−45.

[2] Ferreira HL, Garde R, Fulli G, Kling W, Lopes JP. Characterization of electrical energy storage technologies. Energy 2013;53:288−98.

[3] IEA. Technology roadmap. Energy Storage 2014:37−8.

[4] Zhao H, Wu Q, Hu S, Xu H, Rasmussen CN. Review of energy storage system for wind power integration support. Appl Energy 2015;137:545−53.

[5] Luo X, Wang J, Dooner M, Clarke J. Overview of current development in electrical energy storage technologies and the application potential in power system operation. Appl Energy 2015;137:511−36.

[6] Hameer S, van Niekerk JL. A review of large-scale electrical energy storage. Int J Energy Res 2015;39:1179−95.

[7] Launchpnt. Gravity power grid-scale electricity storage system. 2015 [Online]. Available: http://www.Launchpnt.com.

[8] Rastler D. Electricity energy storage technology options: a white paper primer on applications, costs, and benefits. USA: Electric Power Research Institute (EPRI); 2010. Technical Update.

[9] Punys P, Baublys R, Kasiulis E, Vaisvila A, Pelikan B, Steller J. Assessment of renewable electricity generation by pumped storage power plants in EU Member States. Renew Sustain Energy Rev 2013;26:190−200.

[10] Saini RP. Large-scale energy storage systems. Roorkee: Indian Institute of Technology; 2011.

[11] Caralis G. Hybrid systems (wind with pumped storage). International conference: RES & energy savings in islands, local planning and European cooperation, 23, 24 & 25 October 2009, Milos.

[12] Papaefthymiou SV, Papathanassiou SA. Optimum sizing of wind-pumped storage hybrid power stations in island systems. Renew Energy 2014;64:187−96.

[13] Manolakos D, Papadakis G, Papantonis D, Kyritsis S. A stand-alone photovoltaic power system for remote villages using pumped water energy storage. Energy 2004;29:57−69.

[14] Li Y, Yang M, Zhao L, Wang F. The research of wind-light complementary based on pumped storage power system. Adv Mater Res 2012;354−355:1132−6.

[15] Sciacovelli A, Vecchi A, Ding Y. Liquid air energy storage (LAES) with packed bed cold thermal storage − from component to system level performance through dynamic modelling. Appl Energy 2017;190:84−98.

[16] Locatelli G, Invernizzi DC, Mancini M. Investment and risk appraisal in energy storage systems: a real options approach. Energy 2016;104:114−31.

[17] Pickard WF. The history, present state, and future prospects of underground pumped hydro for massive energy storage. Proc IEEE 2012;100(2):473−83. https://doi.org/10.1109/JPROC.2011.2126030.

[18] Mansoori AG, Enayati N, Agyarko BL. Energy: sources, utilization, legislation, sustainability, Illinois as model state. World Scientific Publishing Co. Pte. Ltd; 2016.

[19] ESA. Sub-surface pumped hydroelectric storage. Available at: http://energystorage.org/energy storage/technologies/sub-surface-pumped-hydroelectric-storage.

[20] Berchum EV. Pumped hydro storage: pressure cavern large-scale energy storage in underground salt caverns [Master's thesis]. Retrieved from TU Delft Repositories. 2014.

[21] Hendriks M. Local hydroelectric energy storage. A feasibility study about a small scale energy storage system combining. Hydropower, gravity power, spring power and air pressure [Master's thesis]. Retrieved from TU Delft Repositories. 2016.

[22] SINTEF. Storage power plant on the seabed. ScienceDaily; 2013. Retrieved January 1, 2017 from: www.sciencedaily.com/releases/2013/05/130515085343.htm.

[23] Slocum AH, Fennell GE, Dundar G, Hodder BG, Meredith JDC, Sager MA. Ocean renewable energy storage (ORES) system: analysis of an undersea energy storage concept. Proc IEEE 2013;101(4):906−24.

[24] Morishige H. Japanese patent 2755778 (expired).

[25] Parker M. PCT patent publication WO 2009/111861 A1.

[26] Walters W. US patent publication 2005/0271501 A1.

[27] Pirkey. US patent 2,962,599.

[28] FAZ. In der tiefe der mere: Hohlkugeln speichern uberschussigen Windstrom-Umwelth & Technik-FAZ. [Online]. Available: https://www.faz.net/aktuell/technik-motor/technik/in-der-tiefe-der-meere-hohlkugeln-speichern-ueberschuessigen-windstrom-1608012.html.

[29] Kibrit B. Pumped hydropower storage in The Netherlands [Master thesis]. Delft: TU Delft; 2013.

[30] Boer WW. The energy island − an inverse pump accumulation station. Arnhem: DNV KEMA Consulting, Lievense BV; 2007.

[31] Kusteilanden. kusteilanden, eilanden voor de kust. 2016. Retrieved from: http://kusteilanden.be/.

[32] Heindl E. Hydraulic rock storage. An innovative energy storage solution for 24 h energy supply. 2013. Retrieved from: http://www.heindl-energy.com/.

[33] Gravity Power. Gravity power module. Energy storage. Grid-scale energy storage. 2011. Available at: http://www.gravitypower.net/.

[34] Oldenmenger WA. Highrise energy storage core: feasibility study for a hydro-electrical pumped energy storage system in a tall building [Master's thesis]. Retrieved from TU Delft Repositories. 2013.

[35] Berrada A, Loudiyi K, Zorkani I. System design and economic performance of gravity storage. J Clean Prod 2017;156:317−32.

[36] Berrada A, Loudiyi K, Zorkani I. Valuation of energy storage in energy and regulation markets. Energy 2016;115: 1109−18. https://doi.org/10.1016/j.energy.2016.09.093.

[37] Berrada A, Loudiyi K, Zorkani I. Profitability, risk, and financial modeling of energy storage in residential and large scale applications. Energy 2017b;119:94−109.

[38] Tarigheh A. Master thesis: gravity power module [Master thesis]. Delft: Delft University of Technology; 2014.

[39] Rehman S, Al-Hadrani, Alam M. Pumped hydro energy storage system: a technological review. Renew Sustain Energy Rev 2015;44:586−98.

[40] EPRI. Electricity energy storage technology options. Electric Power Research Institute; 2010.

[41] Aneke M, Wang M. Energy storage technologies and real life applications − a state of the art review. Appl Energy 2016;179:350−77.

[42] EPRI-DOE. Handbook of energy storage for transmission & distribution applications. Palo Alto, CA; Washington, DC: EPRI; U.S. Department of Energy; 2003.

[43] EPRI. Handbook of energy storage for transmission or distribution applications. Palo Alto, CA: Electric Power Research Institute; 2002. 1007189.

[44] Beaudin M, Zareipour H, Schellenberglabe A, Rosehart W. Energy storage for mitigating the variability of renewable electricity sources: an updated review. Energy Sustain Dev 2010;4:302−14.

[45] Chen H, Cong Y, Yang W, Tan C, Li Y, Ding Y. Progress in electrical energy storage system: a critical review. Prog Nat Sci 2009;19(3):291−312.

[46] Hadjipaschalis I, Poullikkas A, Efthimiou V. Overview of current and future energy storage technologies for electric power applications. Renew Sustain Energy Rev 2009; 13(6−7):1513−22.

[47] Zhou Z, Benbouzid M, Charpentier J, Sciuller F, Tianhao Tang T. A review of energy storage technologies for marine current energy systems. Renew Sustain Energy Rev 2013;18:340−400.

[48] Akinyele DO, Rayudu RK. Review of energy storage technologies for sustainable power networks. Sustain Energy Technol Assess 2014;8:74−91.

[49] Hannan MA, Hoque MM, Mohamed A, Ayob A. Review of energy storage systems for electric vehicle applications: issues and challenges. J Sustain Renew Energy Rev 2017; 69:771−89.

[50] Chatzivasileiadin A, Ampatzi E, Knight L. Characteristics of electrical energy storage technologies and their applications in buildings. Renew Sustain Energy Rev 2013;25:814−30.

[51] Letcher MT. Storing energy, with special references to renewable energy sources. Elsevier; 2016.

CHAPTER 2

Technical Design of Gravity Energy Storage

INTRODUCTION

Energy storage technologies offer a number of services to improve the operation of the utility grid. Such services include energy arbitrage, spinning reserve, regulation, load following, load leveling, and others. However, because of the challenges faced by energy storage, a number of technologies are still under development. Some of these challenges include the optimal design, sizing, and construction of the system. In addition, materials used in the construction of energy storage also play a prominent role in the development of these systems. The high-quality demanding energy storage market has triggered the need for these types of studies to ensure the development of optimal systems [1]. Many researches have carried out studies about the identification of materials for various energy storage systems. An overview about materials used by different types of energy storage including thermal, electrochemical, mechanical, and electromagnetic storage is provided by Liu et al. in Ref. [2]. Whittingham discusses electrical energy storage systems and their material challenges [3]. Thermal energy storage materials are discussed by Fernandez et al. in Ref. [4]. In addition, finite element analysis (FEA) has been performed by several researchers to identify materials suitable for a number of technologies. For instance, the FEA method along with fuzzy decision-making methodology has been used by authors in Ref. [5] to determine the most suitable materials to be applied for biogas storage container. Although, there are a high number of research studies about energy storage materials, few material analyses for gravity energy storage (GES) exist in literature. One of the main objectives of this chapter is to perform a material selection for GES components.

Several researches have investigated methodologies to size energy storage based on different objectives. A generic algorithm has been developed by authors in Ref. [6] to optimally size energy storage with an aim to reduce the operation cost of a microgrid. A similar objective has been studied by Ref. [7] with the use of gray wolf optimization approach. A different model has been developed by Atwa and El-Saadany to size energy storage with a goal to minimize annual energy cost and wind energy curtailment [8]. The sizing and sitting of energy storage has been investigated using a hybrid generic algorithm by Ref. [9]. Maximization of the net present value of a hybrid plant is another objective that has been studied by researches [10,11] for sizing different types of energy storage technologies. Zheng proposed a model to maximize the revenues for a distribution network by sizing energy storage [12]. Authors in Refs. [13–15] developed stochastic algorithms to optimally size storage with respect to the variable energy production and demand profiles. The optimal storage sizing in a distribution network has been determined by Bennett et al. using an energy-scheduling approach, aiming at minimizing peak energy demand [16]. A unit commitment problem solved by a particle swarm optimization model with a goal to reduce the microgrid cost and increase the plant revenues has been studied by Ref. [17]. Energy storage systems can be sized using various types of optimization algorithms as proposed by Refs. [18–21].

The topics covered by this chapter are organized as follows. In Section 2, a conceptual design approach for GES is proposed. This design is improved by an optimization model presented in Section 3. Then, materials that should be applied to the different parts of GES are investigated in Section 4. Modeling and simulation of this storage technology is performed using SolidWorks (SW). FEA is conducted in this chapter to examine the performance and stability of the container under pressure. Section 5 presents a sizing methodology for gravity storage based on the technical and economic aspects of the system. Next, a design improvement for GES with the use of compressed air is discussed and studied. GES improvement using compressed air is studied in Section 6. Finally, the conclusion of this chapter is presented in Section 7.

Gravity Energy Storage. https://doi.org/10.1016/B978-0-12-816717-5.00002-5

DESIGN OF GRAVITY STORAGE COMPONENTS

The identification of GES parameters enables the proper design optimization of system. A methodology to investigate the optimal dimensions of the piston and the container components is discussed in this section.

Energy Equation

The energy expression of GES is

$$E = m_r g z \mu. \tag{2.1}$$

where E stands for energy storage generation (J), m_r is the relative mass of the piston relative to the water, z is height of water in the container (m), g is the gravitational acceleration (m/s^2), and μ is the system efficiency.

This formula can be written in terms of density:

$$E = (\rho_p - \rho_w)\left(\frac{1}{4}\pi D^2 h\right)gz\mu \tag{2.2}$$

where ρ_p, ρ_w are the density of the piston and water, respectively. The piston height is h, whereas its diameter is D. This latter represents also the container's diameter. This expression is a function of the container and piston dimensions. Therefore, it is important to identify the optimal parameters of these components to optimize the system storage capacity. Increasing the height of the piston (h) would lead to a higher energy production, although this would result in a reduction of the water height (z). It is to be noted that a greater water depth would increase the storage capacity. Therefore, it is interesting to investigate the height of the aforementioned parameters, which results in an optimal design. This is found by setting the expression in (Eq. 2.3) equal to zero and solving for "h" to determine the critical points.

$$\frac{\delta E}{\delta h} = 0 \tag{2.3}$$

The sum of the piston and water heights represents the container's height. Therefore, the water depth can be replaced by ($H_C - h$), which represents the container's height minus the piston's height. Solving the above equation (Eq. 2.3) results in an optimal size of the piston height, which should be equal to the water elevation (z). Therefore, both of the studied parameters (h and z) are equivalent to half of the container's height. Replacing the aforementioned parameters in the energy equation results in Eq. (2.4), which is expressed only by the container's dimensions.

$$E = \frac{1}{16}\pi g \mu (\rho_p - \rho_w)D^2 H_c^2 \tag{2.4}$$

There exist a number of sizing possibilities following the desired energy storage capacity. The most appropriate configuration should be selected to optimize the system design. The storage capacity is increased by increasing the container dimensions, which include its diameter and height. Nevertheless, constructing a container with larger diameter may result in piston jamming. Piston jamming occurs if [22]

$$\mu' > \frac{1}{2}\frac{h}{e} \tag{2.5}$$

where μ' is the coefficient of friction related the container's wall and the piston. It is equal to 0.5 for a piston made of steel casing with a concrete container. e is the eccentricity of the load acting on the piston and is equal to Eq. (2.6),

$$e = \frac{D}{2} \tag{2.6}$$

The dimensioning of the piston must strictly consider avoiding piston failure (jamming). This is achieved by restricting its height and diameter accordingly. That is, the height of the piston should be higher than half of its diameter (Eq. 2.7).

$$\frac{h}{D} > \frac{1}{2} \tag{2.7}$$

Parametric Study

A number of parameters have to be identified in the system design study. These include the thickness of the different components such as the container and the return pipe. The container of GES is subject to high-pressure load including both external and internal pressure. Therefore, it should be constructed with an appropriate thickness to resist the total pressure. Lateral earth load represents an external pressure applied to the system; while the piston load and the water pressure act internally on the container structure. The circular wall of the cylindrical container is subject to the hoop tension force, which acts on it. This latter force has to be calculated to determine the thickness of the container [23]. Eq. (2.8) is used to express the maximum hoop tension H_t at the bottom of the structure.

$$H_t = \frac{(\rho_p g + w - k_0 \gamma)HD}{2} \tag{2.8}$$

where k_0 represents the earth pressure coefficient at rest, w is the specific weight of the water, and γ is the soil weight in (kg/m^3).

The stucture stress σ_{st} has to be determined using Eq. (2.9) to ensure that the tensile stress is lower than the permissible stress in concrete [24].

$$\sigma_{st} = \frac{H_t}{1000t + (m-1)A_{st}} \tag{2.9}$$

The modular ratio m is used to calculate the stress in concrete. This latter is calculated using Eq. (2.10). The modular ratio is a function of the compression stress in concrete σ_{cbc}. A_{st} represents the area of steel reinforcement, which depends on the hoop tension and the structure stress (Eq. 2.11).

$$m = \frac{280}{\sigma_{cbc}} \qquad (2.10)$$

$$A_{st} = \frac{H_t}{\sigma_{st}} \qquad (2.11)$$

The return pipe dimensions have to be determined to obtain an overall parametric study of system components. The return pipe represents a connection between the inlet and outlet of the container. Therefore, its height should be equivalent to the container height. In addition, the calculation of the return pipe diameter requires the use of fluid flow formulas. The system discharge rate Q depends on the area and the water velocity. The continuity equation (Eq. 2.12) is used for the identification of the return pipe diameter.

$$Q = V_1 A_1 = V_2 A_2 \qquad (2.12)$$

The average velocity depends on the discharge time of the system, which is the time it takes the piston to move from the top to the bottom of the container. The return pipe thickness is found using Barlow's formula.

The identification of the aforementioned parameters enables the calculation of GES rated power, which is a function of the storage capacity and the discharge time. The equations presented in this section are used to properly design the different components of GES while taking into consideration the technical constraints of the system. The system designed in the parametric study is used as a case study in this book to investigate the validity of the different studied models.

Case study

The storage capacity of GES used in the design study is 20 MWh. The different components of the system are sized according to this capacity. In addition, the container is excavated underground with a depth of 500 m. The different dimensions of the system, including the diameter and thickness of the container and return pipe, are calculated using Eqs. (2.8)–(2.12). A summary of these parameters is presented in Table 2.1.

The system parameters shown in Table 2.1 are used to calculate the storage capacity of GES. The system efficiency has been estimated as 80%. The resulting storage capacity used in this case study has been found

TABLE 2.1
Energy Storage Parameters Used in the Case Study.

	Height (m)	Inner Diameter (m)	Thickness (m)
Container	500	5.21	7.6
Piston	250	5.21	–
Return pipe	500	0.6	0.083

equal to 20 MWh (Eq. 2.4). In addition, the rated power is 5 MW as the discharge time of the system is estimated as 4h.

System Construction Cost

GES is excavated underground for a specific depth. The system capital cost is a function of the excavation cost of the structure. To estimate this latter cost, Eq. (2.13) can be used.

$$E_C = E_V + E_{UC}. \qquad (2.13)$$

where E_C represents the excavation cost, E_V is the excavation volume, and E_{UC} is the excavation unit cost.

The container of the system can be made of reinforced concrete. To approximate the construction cost of the container, Eq. (2.14) is used. This latter depends on the cost of concrete, the cost of the vertical and horizontal reinforcing steel bars, and the cost of the formwork used to build the structure [25,26].

$$C_T = C_c + C_R + C_F \qquad (2.14)$$

Concrete cost is found using Eq. (2.15);

$$C_c = V_c C_{cu} \qquad (2.15)$$

V_c represents the volume of concrete, and C_{cu} is the cost of concrete per unit (€/m³).

$$C_R = V_C \gamma_S C_S (\beta_c + \beta_v) \qquad (2.16)$$

where γ_S is the steel unit weight (t/m³), C_S is the cost of steel per unit (€/t), and β_c is the circumferential steel ratio, which can be estimated using Eq. (2.17)

$$\beta_c = \frac{A_S}{A_C} \qquad (2.17)$$

Here A_S and A_C represent the area of the circumferential reinforcement and the concrete section, respectively. The vertical steel used in the construction of the structure is a ratio of the concrete gross area. It is typically estimated a 1%. In this case study, the structure reinforcement ratio is taken as 1% for the container base and the roof and 2% for the structure walls.

The concrete structure makes use of formwork as mold in the construction process. To calculate the cost of the formwork, Eq. (2.18) can be used.

$$C_F = A_T C_{FU} \qquad (2.18)$$

where C_{FU} represents the formwork cost for a double face unit.

DESIGN MODEL OPTIMIZATION

Reducing the cost of energy storage is one of the most important challenges facing the development of energy storage technologies. Therefore, it is important to optimize the design of the system to obtain a cost-effective device. The cost of GES can be reduced further by optimally dimensioning the different components of the system, particularly the container structure. This section presents a mathematical model that can be used to determine the optimal design of GES based on a specific storage capacity. The developed model has an aim to reduce the construction cost while taking into account the system constraints to prevent its failure. The objective function of the design model is given as:

$$min_c \, f(C) = k_1 D^2 + k_2 D^2 H + k_3 DH + k_4 D + k_5 H + k_6 \qquad (2.19)$$

Subject to:

$$D^2 H^2 - k_7 = 0 \qquad (2.20)$$

$$H - D > 0 \qquad (2.21)$$

$$0 < H \le H_L \qquad (2.22)$$

$$D > 0 \qquad (2.23)$$

The model is formulated as a nonlinear programming (NLP) problem. The objective function minimizes the cost of the system by optimally dimensioning the height and the diameter of the container, the piston, and the return pipe. The construction and excavation costs of the aforementioned components are considered by the objective function f(C) of Eq. (2.19). The system components are sized according to a particular storage capacity. This is represented by the first constraint (Eq. 2.20), which is expressed by only the parameters of the container structure. The height and the diameter of the container should also be constrained by (Eq. 2.7) to prevent piston jam. The excavation of the system should not exceed a specific depth. Therefore, the height of the container should be restricted

TABLE 2.2
k Values Used the Optimization Model.

Parameter	Value
k_1	1,780
k_2	397
k_3	6,887
k_4	38,183
k_5	14,135
k_6	149,748
k_7	6,805,078

to a maximum limitation H_L as shown in (Eq. 2.22). Lastly, the container's parameters (height and diameter) must be greater than zero Eqs. (2.22), (2.23).

A simulation of the proposed model, making use of the case study described before, has been performed in this analysis. The k values of the optimization model have been determined using Eqs. (2.13)–(2.18). k_7 was calculated using the system energy equation presented in Eq. (2.4). Table 2.2 provides a summary of the calculated k values used by the simulation model.

The presented model was implemented and solved using MATLAB. Fig. 2.1 illustrates the results of this design model. The container's height should be equal to 450 m to optimize the design of the system. The diameter of this latter has been found as 5.8 m. The construction and excavation costs of the system are equal to 30.766 M€. This is the minimum cost to construct a 20 MWh system.

To investigate the impact of the container height on GES total cost, an optimization study has been conducted on a system with a lower container's height. This latter has been limited to 100 m. The output of this study demonstrates that the diameter of the container has to be increased to 26 m to achieve a similar energy capacity production (20 MWh). The resulting cost has increased significantly from approximately 30.7 M€ to 48 M€. Therefore, the proposed model demonstrates that it is more profitable to increase the container height rather than increasing its diameter (Fig. 2.2).

STORAGE APPLIED MATERIALS

Properly designing an energy storage system entails taking decisions related to material selection. It is important to appropriately perform a material study to investigate the materials that should be applied to the different components of GES.

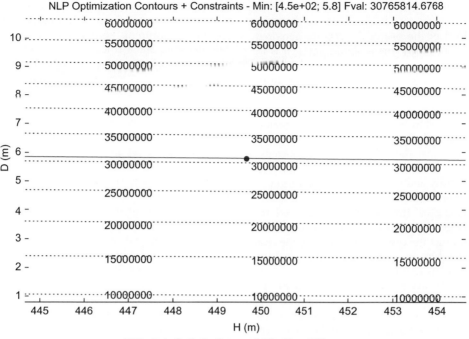

FIG. 2.1 Optimization model for $H_L = 500$ m.

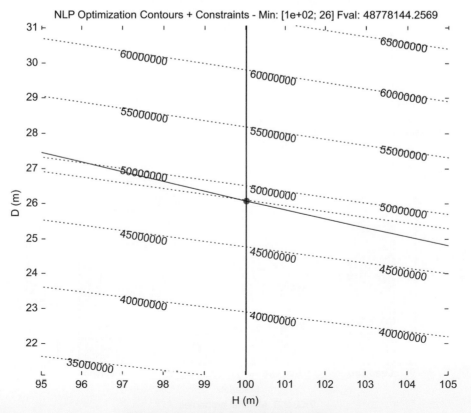

FIG. 2.2 Optimization model for $H_L = 100$ m.

Piston Material

This storage technology makes use of a large piston to operate. The load of the piston delivers the required pressure to the system. Its role is significant as it is in charge of pressurizing the water during the discharging process of GES. Therefore, it should be constructed from suitable materials to respond to the system demanding environment. The characteristics of the different materials that may be used in the construction of the piston should be studied and compared to properly select the most cost-effective one. The two important characteristics used as criteria for the selection of the optimal material include density and cost. Eq. (2.2) demonstrates that the energy production of GES changes in accordance with a change of the piston density. Therefore, the system capacity depends on the density of the material used to construct the piston. Furthermore, the cost of the material is also an important criterion, which would result in an economic system.

The investigated inner materials of the piston used in this study include concrete, lead ore, aluminum, and iron ore. The density and cost of the aforementioned materials are illustrated in Fig. 2.3. A number of aspects were not included in the presented costs such as transportation, material processing, and residual value. The consideration of these aspects would result in higher construction cost [27].

The material with the highest density is lead ore as shown in Fig. 2.3. However, this material is more expensive. Therefore, it is interesting to compare the energy production of the system versus the construction cost of the piston; Fig. 2.4 illustrates this comparison. It is shown that constructing the piston with lead ore would drastically increase the system cost compared with other investigated materials. Hence, it should not be chosen as an optimal material due to its uneconomic viability. Although, aluminum and concrete have lower costs, their densities are also low. The density of iron represents the double of concrete density. Therefore, even if concrete material is less expensive than iron ore, constructing the piston with concrete would result in a higher cost. The reason behind this relies on the fact that a larger concrete piston would be needed to counterbalance the low density of concrete. Fig. 2.5 illustrates that concrete piston with a double diameter should be used to produce the same energy production. This diameter has to be doubled for concrete piston to reach the same iron ore piston mass. Nevertheless, the construction cost of the container structure would be increased if the diameter of the piston is enlarged. Therefore, compared with its counterparts, iron ore is considered an optimal material which can be used for the construction of the inner structure of the piston because of its low cost and interesting density.

To prevent the chemical mixing of iron ore with water, the piston inner structure has to be cased. As the load of the inner piston is high, the casing used should have high strength. Steel is a suitable material to construct the outer structure of the piston due to its robustness and water tightness. In addition, to avoid the flowing of water across the piston and reduce leakage, it is crucial to use hydraulic seals. These seals should be made of materials able to resist the high pressure within the system. Materials commonly used in hydraulic applications include plastic polymers such as polytetrafluoroethylene. This material has high wear resistance and minimizes the friction between the piston casing and the container.

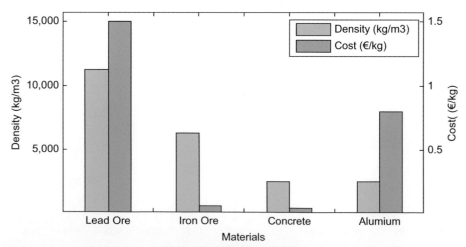

FIG. 2.3 Material properties.

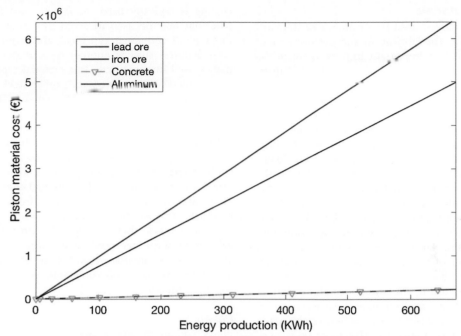

FIG. 2.4 Piston cost based on different materials and capacity.

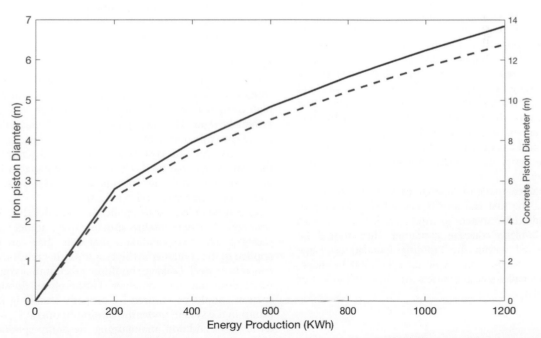

FIG. 2.5 Comparison of energy produced by concrete and iron piston.

Container Materials

Containers have been used in a number of domains and applications. The fabrication and performance of containers have seen important improvement in the last years. Choosing the proper materials for constructing the container requires a material study while considering a number of aspects. The most important criteria used in the selection of the container construction material include robustness, durability, and cost. The container structure should withstand the high pressure applied by the water and the piston loads. In addition, the storage containers should have the strength to resist the internal/exterior pressure of the medium. They must resist to corrosion and be watertight. Aboveground containers should be able to withstand all weather conditions. That is, they should be weather resistant. Underground containers should resist the weight of the soil surrounding it and any other pressure that might occur at the top the container. In addition, the surface of the container structure should be smooth to reduce friction between the container and the piston. Containers can be made of various types of materials. The most used ones include the following:

Concrete

Concrete containers are durable and can last for a number of years. Materials such as plasticizers can be added to concrete to augment its strength. In addition, concrete is put into a seamless mold to avoid leakage. One of the drawbacks of concrete containers is that they are not repaired easily as compared with steel containers. They are susceptible for tensile stresses and should be reinforced to increase its resistance. The durability of the container may be affected by steel reinforcement as it may cause the structure to crack due to steel corrosion. In addition, to reduce friction, it is necessary to smooth the surface of concrete containers.

Containers made of concrete may be precast or cast-in-place. The first category is cost-effective in sizes of thousands cubic meters or more. To decrease the cost of cast-in-place concrete container, they should be incorporated within the building foundations. Large concrete containers are typically cylindrical in shape, whereas smaller size containers are either cylindrical or rectangular.

Steel

Containers made of galvanized steel corrode over time and rusts. To protect them against corrosion, a zinc coating is typically used. Aquaplate steel is another type that may be used to avoid this issue. However, this type of containers can break and requires a high cost. Stainless steel is another type of steel material that is used in containers. This material type is much more expensive compared with other steel materials. Its tensile resistance is high. This latter is one of the important characteristics required by GES. Furthermore, it has a surface that is relatively smooth. Steel containers with large capacities are usually cylindrical in shape. Large aboveground tanks can hold up to several million cubic meters. These latter already exist and are used for a number of applications such as thermal energy storage. To protect the interior of the tank from corrosion, materials such as epoxy coating are used. For small-scale steel tanks, they are typically constructed from galvanized sheet steel and have a rectangular shape. There capacities are lower than 100 m^3. Steel containers are more expensive and vulnerable to corrosion than concrete containers.

Plastic

Plastic containers can be used for small-scale applications. They are typically found as prefabricated modular units. Aboveground tanks needs to be protected against sunlight UV radiation using opaque covering or stabilizers. One of the advantages of plastic tanks is their light weight. The most common type of plastic containers is polyethylene. It is light, noncorrosive and has high strength and low cost.

Fiberglass

Less commonly used material for the construction of containers is fiberglass. Their resistance to corrosion is high compared with other materials although they are extremely rigid. They are quite brittle and may crack as they are thin and light.

The construction cost of the container depends on the type of material used. The cheapest manufacturing materials are plastic and fiberglass followed by steal and precast reinforced concrete [28].

GES used in large-scale application requires the construction of a large container structure that is under high pressure. The two candidate materials that can be applied to the container in this case are either reinforced concrete or steel. Leakage should be taken into account while designing the container. Furthermore, flooded suction should be ensured by properly installing the pump in the system under the minimum operating water level. In addition, maintaining the pump specified net positive suction pressure is important to prevent subatmospheric pressure conditions.

Reinforced concrete container offers a number of advantages over steel container. These include long lifetime of about 50 years and good resistance to compression stresses [29]. Other issues faced by steel containers include buckling, corrosion, and geometric imperfections. Conversely, high tension resistance and leak-free configuration are expected from steel containers. Containers made of reinforced precast concrete are also subject to a number of challenges such as large wall thickness and low tensile strength. This type of containers is commonly used because of their cost-effectiveness compared with other types of containers [30]. The case studies used in this book for large-scale application make use of a reinforced concrete container because of the different advantages it offers over its counterparts.

To examine the impact of pressure on the aforementioned studied materials, a FEA is carried out in this section.

Selection of Material Using Finite Element Analysis Method

To investigate the impact of the pressure loads on the container structure, a simulation study is performed on the container/piston assembly while taking into consideration the boundary conditions. The stress, displacement, and strain of the different studied container materials are compared using FEA. A case study has been performed on a laboratory prototype designed using SW for the different applied materials. The height and the diameter of the designed container are 1.5 and 0.5 m, respectively.

SW FEA program is used to perform a FEA to examine the response of the pressurized container. In this simulation, the four investigated materials will be applied to the structure of the container, which is under external and internal loads. As the water pressure increases with depth, a varying internal pressure is applied to the container. The piston load applies an additional pressure on the structure. In addition, a cyclic load is also experienced when the container is emptied and filled. The container is subjected to a total internal pressure expressed in Eq. (2.24); where x represents the pressure at a specific location.

$$P_x^i = \rho_w g x_H + (\rho_p - \rho_w)gh. \tag{2.24}$$

Underground GES is also subject to lateral earth pressure. This external load is given by Eq. (2.25):

$$P_x^e = k_0 y x_H. \tag{2.25}$$

The structure stability of the container is analyzed by applying the aforementioned loads on the walls and the bottom of the container. The piston applies pressure on water rather than on the container. Hence, the piston load is represented by horizontal and vertical pressure on the water. In addition, the distribution of the water load applied on the container is nonuniform as shown by the first part of Eq. (2.24). To compare the structure stability of the container made from different materials, four simulations would be performed in this case study. The outcomes of these simulations are shown in Figs. 2.6–2.8.

The container strength is the stress at which the structure will start to fail. The structure failure is analyzed by SW using maximum-distortion-energy theory (von Mises). The output of von Mises equivalent stress is shown in Fig. 2.6. Changes in stress within the container is shown through changes in colors from blue to red. An increase of stress inside the structure is indicated in this case (Fig 2.6). Maximum stress has been reached near the outlet of the container for the four simulated scenarios due to the significant pressure applied there. The maximum value varies from one material to the other. To examine the yielding of the structure, it is important to compare the material's yield strength with the obtained von Mises equivalent stress. The factor of safety (FOS) of the structure has to be calculated and compared with the required FOS. This is found by dividing the material yield strength over the resulting maximum von Mises stress. In this case study, the outcomes of the calculated FOS have shown that the lowest FOS has been obtained by reinforced concrete container and is equal to 2.5. The minimum FOS required by various codes is in the range of 1.5–3.0. Therefore, the results of this analysis demonstrate that the container structure made of the different investigated materials is not expected to fail under the applied pressure.

A displacement and strain analysis are also performed by SW to investigate the structure failure. Fig. 2.7 illustrates the outcomes of the displacement analysis. The bottom of the container experiences the highest material displacement due to the high pressure occurring at the bottom. This high pressure is caused by the piston and the water loads. The results of this analysis demonstrate that the plastic container has the highest material displacement, which is equivalent to 3.207 mm. This is followed by concrete, glass, and steel container. Therefore, pressure has less impact on steel container compared with their counterparts.

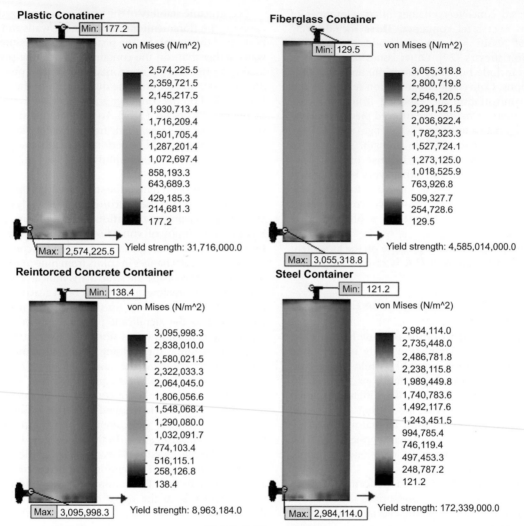

Plastic Conatiner

Min: 177.2

von Mises (N/m^2)

2,574,225.5
2,359,721.5
2,145,217.5
1,930,713.4
1,716,209.4
1,501,705.4
1,287,201.4
1,072,697.4
858,193.3
643,689.3
429,185.3
214,681.3
177.2

Max: 2,574,225.5   Yield strength: 31,716,000.0

Fiberglass Container

Min: 129.5

von Mises (N/m^2)

3,055,318.8
2,800,719.8
2,546,120.5
2,291,521.5
2,036,922.4
1,782,323.3
1,527,724.1
1,273,125.0
1,018,525.9
763,926.8
509,327.7
254,728.6
129.5

Yield strength: 4,585,014,000.0

Max: 3,055,318.8

Reintorced Concrete Container

Min: 138.4

von Mises (N/m^2)

3,095,998.3
2,838,010.0
2,580,021.5
2,322,033.3
2,064,045.0
1,806,056.6
1,548,068.4
1,290,080.0
1,032,091.7
774,103.4
516,115.1
258,126.8
138.4

Max: 3,095,998.3   Yield strength: 8,963,184.0

Steel Container

Min: 121.2

von Mises (N/m^2)

2,984,114.0
2,735,448.0
2,486,781.8
2,238,115.8
1,989,449.8
1,740,783.6
1,492,117.6
1,243,451.5
994,785.4
746,119.4
497,453.3
248,787.2
121.2

Max: 2,984,114.0   Yield strength: 172,339,000.0

FIG. 2.6 Stress analysis.

As for the strain analysis, the simulation outcomes are presented in Fig. 2.8. High strain is experienced by the plastic container as shown by the red color spread throughout the structure. The maximum strain value is 3.073×10^{-3} for plastic container. The strain outcomes are low for steel, concrete, and fiberglass.

Based on the results obtained from the performed structure failure analysis, a container made of steel demonstrates good prospects compared with its counterparts. However, reinforced concrete containers should also be used.

SIZING OF GRAVITY STORAGE SYSTEM

Matching energy demand with power production is one of the most important challenges facing the development of renewable energy storage. To overcome this issue, the installation of energy storage systems is crucial as they enable the balancing of renewable energy generation with load demand. Optimally sizing the storage system used is necessary to effectively and economically operate the hybrid renewable plant. The purpose of this section is to investigate the optimal sizing of GES installed in a renewable energy plant.

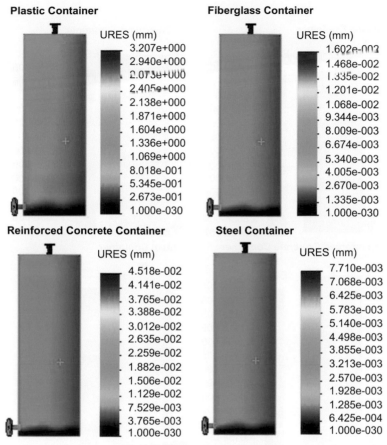

FIG. 2.7 Displacement analysis.

The proposed model optimally size GES with regard to the system-generated profit. As GES has a scalable capacity, the system can be sized according to the plant requirement while maximizing its revenues. The mathematical sizing model is subject to the storage system technical characteristics and plant operation.

The aim of this problem is to minimize the storage installation cost. This is achieved by optimally dispatching energy between the different components of the hybrid renewable energy plant and the electric grid. Short-term forecast of load profiles and energy production are used as input data. Assumptions used by the model are similar to those of the operation optimization model proposed by Ref. [31]. Energy transactions take place on the day-ahead market, with hourly energy dispatch for a 24 h period. The model takes into the transmission capacity of the electric grid. On an hourly basis, the performance of the storage system and energy

output are assumed to be constant. In addition, instantaneous storage start-up is considered. Lastly, energy prices are presumed to remain constant.

The hybrid renewable energy plant consists of a wind farm connected to GES. The hybrid plant is linked to the utility grid as shown in Fig. 2.9.

Sizing Algorithm

The proposed sizing methodology has been formulated as NLP model. The parameters used as inputs by the sizing model include hourly energy prices, RE production profiles, cost, and characteristics of the storage system. The model outputs include revenues and profit of RE plant, energy dispatch profiles, and maximum viable storage capacity. The maximum profit that can be obtained from the use of energy storage in a renewable energy plant is found by altering the storage capacity. This maximum value of this latter is achieved once the profit

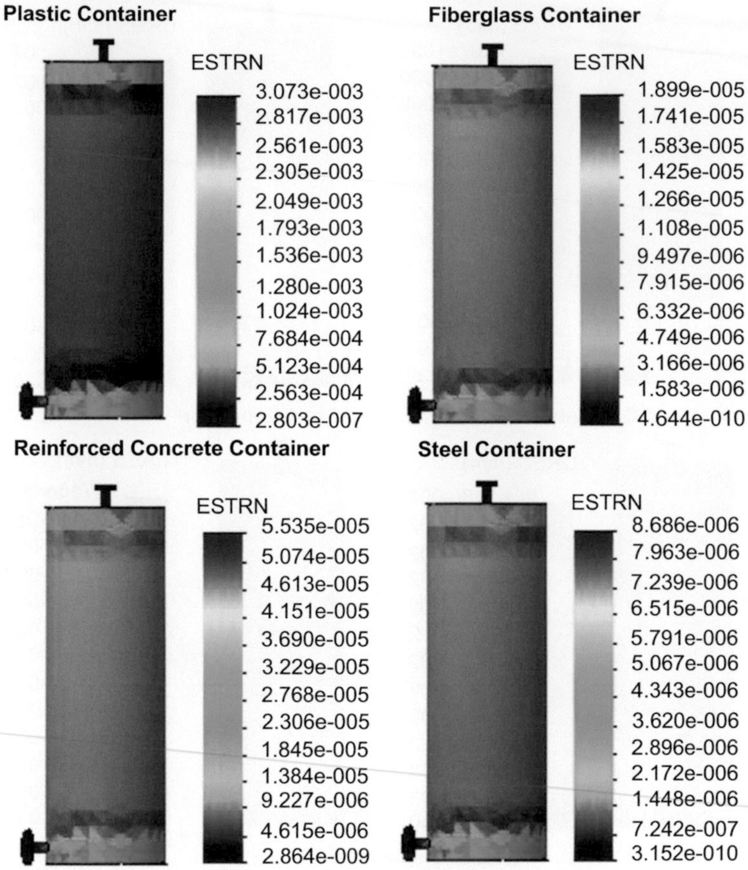

Plastic Container
ESTRN
3.073e-003
2.817e-003
2.561e-003
2.305e-003
2.049e-003
1.793e-003
1.536e-003
1.280e-003
1.024e-003
7.684e-004
5.123e-004
2.563e-004
2.803e-007
Fiberglass Container
ESTRN
1.899e-005
1.741e-005
1.583e-005
1.425e-005
1.266e-005
1.108e-005
9.497e-006
7.915e-006
6.332e-006
4.749e-006
3.166e-006
1.583e-006
4.644e-010
Reinforced Concrete Container
ESTRN
5.535e-005
5.074e-005
4.613e-005
4.151e-005
3.690e-005
3.229e-005
2.768e-005
2.306e-005
1.845e-005
1.384e-005
9.227e-006
4.615e-006
2.864e-009
Steel Container
ESTRN
8.686e-006
7.963e-006
7.239e-006
6.515e-006
5.791e-006
5.067e-006
4.343e-006
3.620e-006
2.896e-006
2.172e-006
1.448e-006
7.242e-007
3.152e-010

FIG. 2.8 Strain analysis.

Renewable Energy Farm
E_C
E_I
E_T
E_T
Transmission Line
Utility Grid
E_S
E_D
Energy Storage

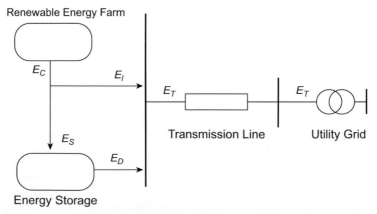

FIG. 2.9 Schematic of the hybrid system.

is maximized. The model objective function is to maximize the profit by properly sizing the storage system. This objective function is expressed as:

$$Max\left[P_r = Re\,v(t) - Cos\,t(t)\right]. \tag{2.26}$$

where P_r is the profit made by the plant owner, $Rev(t)$ the hourly generated revenues, and $Cost(t)$ is the operating cost of the hybrid plant. The aim of this model is to maximize the plan profit, which is calculated by subtracting the plant service costs from the operation revenues.

Hourly revenues generated by the hybrid farm are the product of energy price $P_E(t)$ and energy traded in the energy market (sold to the grid). This energy traded with the grid is the sum of storage discharged energy $E_D(t)$ and energy exchanged between the plant and the grid $E_I(t)$. The revenues equation can be formulated as:

$$Re\,v(t) = [E_D(t) + E_I(t)]P_E(t). \tag{2.27}$$

The cost of an energy storage system consists of fixed costs and variable costs. Fixed costs are paid irrespective of the energy produced by the plant. On the contrary, variable costs are a function of the energy stored in the system. Short-term decisions of energy dispatch require taking into consideration only variable costs. As mentioned by Refs. [32,33], short-term costs consisting of operation and maintenance, fuel, environmental emissions, and start-up costs should be considered by unit commitment and short-term scheduling problems. For GES, there is no cost associated with fuel and emissions as this system does not consume fuel nor produce emission during its operation. Consequently, GES service costs are equal to operation and maintenance costs (Eq. 2.28):

$$Cos\,t(t) = C_{O\&M}\sum(E_S(t)). \tag{2.28}$$

Here $(C_{O\&M})$ is the operation and maintenance cost of the storage system in ($€$/kWh); $E_s(t)$ is the system stored energy.

The energy level of the storage system changes in time in accordance with the energy stored and discharged from the system at that time. The storage state depends also on the system characteristics such the storage self-discharge (δ) and efficiency (η). The energy level at a particular time is found using Eq. (2.29).

$$S(t) = (1 - \delta)S(t-1) - E_D(t) + (E_S(t)\eta). \tag{2.29}$$

The storage energy level ($S(t)$), at a specific time (t), is a function of the storage state at a prior time ($t-1$), as well as the energy added and discharged from the system at that particular time (t). In addition, energy losses should be taken into consideration as shown in Eq. (2.29).

The model is subject to the storage technical characteristics. These include the system rated power and capacity limits. Therefore, the storage energy level should be constrained by Eqs. (2.30), (2.31).

$$0 \leq S(t) \leq S_{Limit}(t). \tag{2.30}$$

The energy level in the system must be greater than zero and lower than the storage capacity limit. In addition, the storage state S(t) should be greater or equal to the energy discharged from the system at that specific time (Eq. 2.30).

$$S(t) \geq E_D(t). \tag{2.31}$$

The energy that flows in and out, at a particular time, from the energy storage system is controlled by Eqs. (2.32), (2.33); where E_L represents the energy storage limit.

$$E_D(t) \leq E_L. \tag{2.32}$$

$$E_S(t) \leq E_L. \tag{2.33}$$

Energy dispatched between the storage system, the RE plant, and the grid must be positive.

$$E_S(t),\,E_D(t),\,E_I(t),\,E_G(t) \geq 0. \tag{2.34}$$

where $E_G(t)$ represents the energy produced by the renewable energy plant.

Energy produced by RE plant is either injected directly to the electric grid or stored in GES system. This model constraint is given by:

$$E_G(t) = E_S(t) + E_I(t). \tag{2.35}$$

Maximization of the plant profit is one of the main objectives of the proposed model. This is done through optimally controlling the charging and discharging of the storage system. A constraint controlling the dispatch of energy should be incorporated in the model to maximize profit. The efficiency of GES is in the range of 80%. There is an energy loss in the process of charging and discharging the system. Therefore, energy should not be stored and discharged simultaneously from the storage system to avoid the loss of energy. Consequently, it is more practical and economical to send energy directly to the utility grid; instead of charging it and discharging it in the storage system.

This constraint is expressed through Eq. (2.36).

$$E_S(t)E_D(t) = 0. \qquad (2.36)$$

Case Study

The effectiveness of the presented sizing model is examined by a case study. General Algebraic Modeling System (GAMS) program is used to solve the proposed NLP model. This model is not limited to sizing GES; it can be used for several types of energy storage. The performed case study makes use of large-scale GES. The characteristics of this system has been used by the model. An efficiency of 80% has been assumed [34]. Input data concerning wind energy production and energy prices were taken from historic data published by "Red Eléctrica de España, S.A." [35]. There exist about 992 wind farms in Spain. Fig. 2.10A shows hourly wind energy production used in this case study. It is shown that maximum energy generation is obtained at night around 11 p.m.; while energy generation is at its minimum value at noon. Hourly electricity prices are presented in Fig. 2.10B. Low energy prices occur at early morning from 4 to 7 a.m. They vary significantly throughout the day and reach their maximum value in the evening especially during peak energy demand periods.

Profit is maximized with the utilization of the optimal energy dispatch strategy by the hybrid

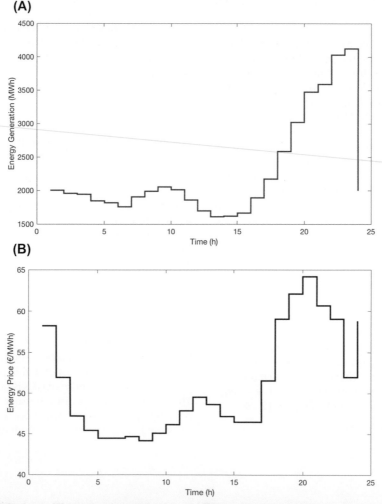

FIG. 2.10 **(A)** Hourly wind production and **(B)** energy prices for the simulated day.

renewable energy plant. At the maximization point, increasing the storage capacity of the system does not affect the profit as it remains constant. Therefore, the plant owner will not take advantage from installing an energy storage system with larger storage capacity. The simulation output the maximum valuable storage capacity, which is obtained when profit, is maximized. Fig. 2.11 shows the daily profit that can be obtained by the plant owner for different capacities of GES. With higher capacities, more profit would be gained. However, beyond a storage capacity of 121 GWh, the profit would remain constant. Therefore, for the presented case study, the maximum sizing capacity of energy storage is 121 GWh, which can be installed as decentralized storage for the different studied wind

plants. It is important to keep in mind that the proposed sizing methodology does not take into consideration the demand side and the transmission capacity constraints [31].

The optimal dispatch of energy for the storage system is determined based on the objective function of the model, which is to maximize profit. The optimal operation of the hybrid plant is to store energy during periods of low energy prices; energy is charged up to the storage capacity limit. When these prices increase, during peak periods, energy is discharged from the storage system to maximize the plant revenues. Fig. 2.12 illustrates the storing of energy in the system. Energy is stored in early morning between 5 and 8 a.m. After the charging period, the system is on standby. At 7

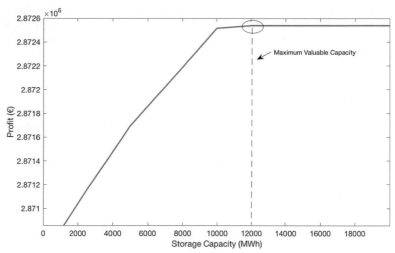

FIG. 2.11 Storage capacity versus profit.

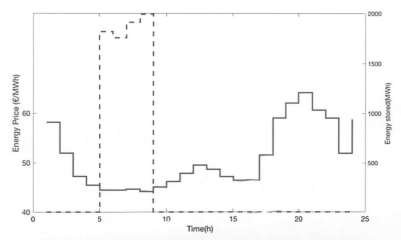

FIG. 2.12 Hourly energy prices versus energy stored in the system.

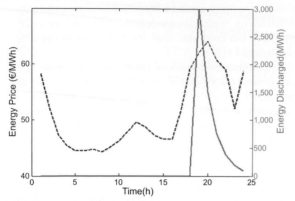

FIG. 2.13 Hourly energy price versus storage discharged energy.

p.m., energy prices go up causing the storage system to start discharging energy until midnight. The discharged energy is sold with higher prices to the utility grid. This will enable the plant owner to increase his revenues and hence maximize his profit (see Fig. 2.13).

Fig. 2.14 shows the generated hourly profit versus the energy prices. It is shown that there is an interesting correlation between these two variables. As the energy prices increase, the profit is increased. During periods of low energy prices, a zero profit is obtained due to the absence of energy trading between the hybrid plant and the utility grid.

The utilization of the proposed model will optimally size the storage system and operate the hybrid plant in an optimal manner with an aim to maximize the plant

profit. Therefore, under the stated assumptions and with respect to the plant restrictions, the maximum valuable storage capacity can be determined using the presented model.

SYSTEM IMPROVEMENT

A challenging issue facing the development of renewable energy is the storing of large amount of energy. There exist only few energy storage options used in large-scale applications. Batteries used in this latter application include lead-acid, sodium-nickel chloride, sodium-sulfur, and nickel-cadmium [36]. Lithium-ion batteries are more suitable for cell phones and electric vehicles. The two prominent systems used in large-scale applications are pumped hydrostorage (PHS) and compressed air energy storage (CAES) [37]. These aforementioned technologies suffer from a number of difficulties such as the geographic site limitation for PHS and the moderate system efficiency for CAES. GES is considered an alternative option that is able to overcome the challenges encountered by these latter systems. To increase the energy production of GES used in large-scale applications, this section proposes the combination of compressed air and GES.

A number of innovative concepts have been studied by researches such as compressed air pumped hydro energy storage, which combines the working principle of CAES and PHS [38]. The proposed system is comprised of a reservoir, a water/air tank, a pump-turbine, a compressor, and a motor-generator. Before the cycle starts, air is compressed in the tank by the air compressor. During the storage charging process, the

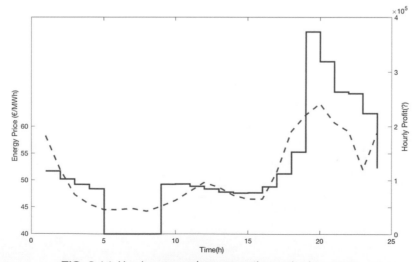

FIG. 2.14 Hourly energy prices versus the received revenues.

pumped water adds more pressure to the compressed air. In this system, air is combined with water in the same tank. A similar system with all the aforementioned components has been studied by Ref. [39]. The only difference between these two systems resides in the configuration of the pressure tank. This latter system does not combine air with water in the same pressure tank. The performance of this system has been analyzed by Wang et al. [40].

Although a number of research papers have tackled the topic of combining the working principle of CAES and PHS, there is a limited work investigating the combination of compressed air with GES. Therefore, it is interesting to study the performance and the design of this latter system.

The physical model of compressed air gravity energy storage (CAGES) and its working principle is presented in this section. An optimal configuration of the system is investigated and discussed. In addition, the proposed system is compared with CAES and PHS to derive its offered benefits.

Description of Compressed Air Gravity Energy Storage

The most commonly used energy storage systems for large-scale application are pumped hydro energy storage and CAES. PHS uses a simple principle to store and discharge energy between its two reservoirs located in different altitudes. During periods of low energy demand, electricity is used to pump water from the lower reservoir to the upper one. When this stored energy is needed, water is transferred from the upper reservoir to the lower one passing through a turbine and a generator, which produces energy. Further development of this storage technology is limited due to its geologic condition, which is not always available. CAES is considered an alternative to pumped hydrosystem as it demonstrates good technical and economic prospects for bulk application. Air is compressed by excess energy and stored in underground cavern. To generate electricity, the stored air is combusted in a gas turbine.

GES system has been proposed as an alternative to pumped hydro energy storage. Toward the improvement of this storage technology, a novel system that combines the working principles of both PHS and CAES is suggested. This system, which is investigated in this chapter, is known as CAGES. The purpose of adding compressed air to gravity storage is to increase the system pressure. Therefore, an air pressure along with a pressure vessel will be integrated to the existing components of GES. The addition of compressed air is

similar to the increase of the difference in height between the two reservoirs of PHS. That is, the increase of water pressure in CAGES. For example, air pressure of 10 MPa added to the system is equivalent to extra 1000 m for raising the water in PHS. Therefore, compressed air is used for the storing of energy in compressed air gravity storage.

A number of challenges faced by CAES and PHS in large-scale application can be overcome with the use of CAGES. Such issues include the high height difference required by PHS and the utilization of air turboexpander, fossil fuel, and gas turbines in CAES. With regard to reliability, CAGES is more efficient as it makes use of hydroturbines.

Design Model

The mechanical equipment of the system consists of a gas compressor, a high-pressure vessel, a reversible pump-turbine, and a motor-generator. Fig. 2.15 shows a schematic of the system. The gas compressor compresses air before the storage phase. The pressure vessel has a volume (V_1) and an initial pressure (P_1). During the charging stage of the system, water is pumped in the container using surplus energy. This

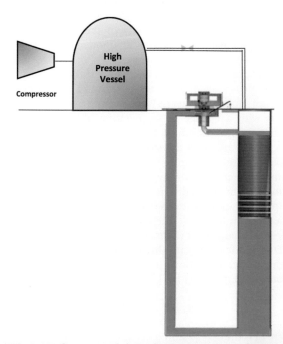

FIG. 2.15 Compressed air gravity energy storage schematic.

pumped water compresses air further leading to a decrease in the container air volume (V_2). This process continues until the air pressure is equivalent to the container withstand pressure (P_2). To produce energy in the generation phase, the piston goes down under the effect of gravity and air/water pressure. The piston downward motion makes water flow in the opposite direction driving a hydroturbine connected to a generator.

Compressing air to the withstand pressure (P_2) value maximizes the storage capacity of the system. This latter depends also on the volume of the compressed air (V_2). It is important to investigate the optimal value of the water-to-air ratio to determine the air volume. In addition, it is also crucial to study the coloration between the aforementioned parameters with energy production to identify the optimal configuration of CAGES.

Energy stored in CAGES system could be categorized in two forms including compressed air and hydraulic form. Therefore, the energy storage capacity of this system is the sum of energy stored in the hydraulic components (E_h) and the compressed air (E_a). Compressed air gravity storage energy equation is shown in Eq. (2.37):

$$E_t = E_a + E_h \qquad (2.37)$$

A similar system, known as compressed air pumped hydro energy storage, has been studied by authors in Ref. [41]. The optimum capacity of this system was determined using mathematical models for both adiabatic and isothermal processes.

A compression process is isothermal if the compression of air is slow and the temperature of the system does not change. Conversely, if a rapid compression occurs with a change in temperature, the process is adiabatic. Compression of air in CAGES occurs before the charging mode of the storage system. Then, at the storage phase, this air is compressed further by the pumped water. This air compression occurs gradually in the system. In addition, a small change of water temperate should be expected (below 5°C) as shown by authors in Ref. [40]. Therefore, the air compression process is assumed isothermal for CAGES system.

During the energy production mode and using an isothermal compression, the energy storage capacity is found as Eq. (2.38) [41]:

$$E_a = \int_{V_2}^{V_1} P_x dV_x \qquad (2.38)$$

where E_a is the energy stored in compressed air, V_x is the air volume, and P_x is the pressure of air in the container. The energy production mode is over when all the water

in the container is discharged. The released energy can be expressed as Eq. (2.39):

$$E_a = P_2 V_2 \ln\left(\frac{V_1}{V_2}\right) \qquad (2.39)$$

Eq. (2.39) should be derived and set equal to zero, to determine the maximum energy which could be released from a pressure vessel with a withstand pressure P_2 and a volume capacity of V_1. The obtained solution enables us to identify the system parameters including the air volume in the storage generation phase and the preset pressure. The aforementioned parameters are given by:

$$\begin{cases} P_1 = \dfrac{P_2}{k_c} \\[2mm] V_2 = \dfrac{V_1}{k_c} \end{cases} \qquad (2.40)$$

The preset pressure (P_1) is obtained from the container maximum pressure and constant k_c. This latter equals to 2.72 as determined by authors in Ref. [41]. Similarly, the volume V_2, which represents the air final volume when the pressure reaches its maximum pressure P_2 is derived from V_1 and k_c.

The energy released from compressed air is found by multiplying the container withstand pressure with its final volume. Therefore, the maximum delivered energy can be expressed as:

$$E_a = P_2 \frac{V_1}{k_c} \qquad (2.41)$$

Energy stored in the hydraulic form is expressed using the energy generation equation of GES without the inclusion of compressed air (Eq. 2.42).

$$E_h = (\rho_p - \rho_w)\left(\frac{1}{4}\pi d^2 h\right) gz\eta \qquad (2.42)$$

E_h is the energy stored in the hydraulic components of CAGES. This energy is a function of the water and piston density (ρ_p and ρ_w); it depends also on the dimensions of the container (height h, and diameter d), as well as the efficiency of the system (η), the elevation height (z), and the gravitational acceleration g.

The energy equation of CAGES is expression as:

$$E_t = P_2 \frac{V_1}{k_c} + \left[(\rho_p - \rho_w)\left(\frac{1}{4}\pi d^2 h\right) gz\mu\right] \qquad (2.43)$$

Simulation

A simulation of the proposed model has been conducted to study the performance of CAGES system. The case

TABLE 2.3	
Case Study System Specifications.	
System Specifications	**Value**
Container height	500 m
Container diameter	5.21 m
Piston height	250 m
Withstand pressure	10 MPa

study parameters are presented in Table 2.3. The piston height has been selected as half the container height to optimize the energy storage capacity. The withstand pressure of system is equivalent to 10 MPa.

To properly design a CAGES system, it is important to determine the optimal air-to-water ratio in the container. The storage capacity of the system is also dependent on the volume of the tank and its pressure. Therefore, it is interesting to analyze the impact of these parameters on the energy production of the system. Furthermore, the energy storage capacity of GES is compared with that of CAES to investigate the added value of compressed air in the system.

Impact of Water-Air Ratio on the System Storage Capacity

To improve the performance of compressed air gravity storage, it is crucial to design a system with maximum air-to-water ratio. Fig. 2.16 shows the storage energy delivered versus the air-to-water ratio. The system energy production increases as this ratio increases. Maximizing this ratio results in an optimum storage capacity. A ratio of 1 is equivalent to an energy production of 23 MWh. The storage capacity is optimized when compressed air constitutes half of the container's volume. The other half of the container is occupied by the piston. A schematic of the optimal system design is shown in Fig. 2.17. It is shown that the upper side of container is not linked to the return pipe. The diameter of this latter should be doubled to hold the water that flows in during the discharging phase of the storage.

A potential capacity can be stored in CAGES as shown in Fig. 2.17. The actual energy production of GES is 19.88 MWh. With the incorporation of compressed air, an amount of 3.22 MWh is added to the system. The total energy production of CAGES system would be equal to 23 MWh as shown in Fig. 2.16.

Impact of the Container Height on the System Storage Capacity

The effect of the container's height variation on the production capacity of the system should be studied. For GES, the height of the container significantly affects the energy production. However, increasing this height would results in an increase in the cost of excavation. This latter would lead to a higher construction cost.

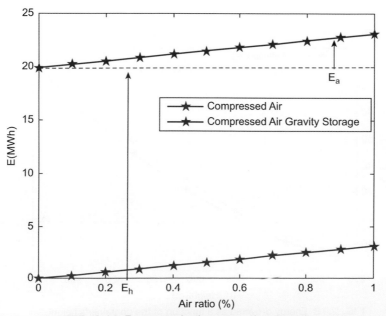

FIG. 2.16 Energy production versus air-water ratio.

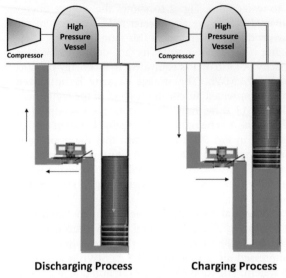

Discharging Process   **Charging Process**

FIG. 2.17 Compressed air gravity storage optimal system.

Fig. 2.18 demonstrates that the container's height has a significant impact on the storage capacity. Larger containers result in higher energy production systems. An important increase is shown for CAES. The addition of compressed to GES increases the water pressure, which is similar to increasing the container's height. Therefore, compressed air could counterbalance the height of the container in GES.

As shown in Fig. 2.18, compressed air could be added to GES to increase the system energy production instead of increasing the height of the container for a specific energy storage capacity. For instance, GES with a capacity of 19.88 MWh requires a container's height of 500 m for a diameter of 5.21 m. To obtain the same energy production for a reduced container's height of 450 m, compressed air should be incorporated in the system. Therefore, the addition of compressed air to this studied system decreases the container's height by 50 m. Combining compressed air with GES results in an improved and optimal energy production.

Impact of Container Pressure and Diameter on the System Storage Capacity

Energy delivered by CAGES is influenced by the maximum withstand pressure of the container as shown by Fig. 2.19. As the system pressure is increased, an upward trend of energy storage capacity is shown. For a pressure of 20 Mpa, the storage produced energy reaches 26.21 MWh. Comparing this latter value with the energy obtained from GES without the incorporation of compressed air, it can be noticed that an interesting amount of energy is added to the system with the addition of compressed air. Consequently, energy storage capacity of CAGES is considerably enhanced with the increase of the system allowable pressure.

The energy that can be stored in CAGES is dependent on the volume and the pressure of compressed air. These latter are a function of the container size,

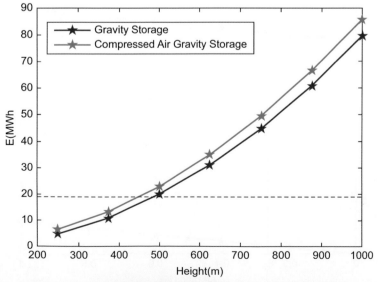

FIG. 2.18 Energy delivered as a function of the container's height.

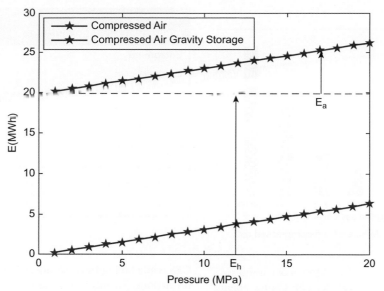

FIG. 2.19 Energy delivered versus the system withstand pressure.

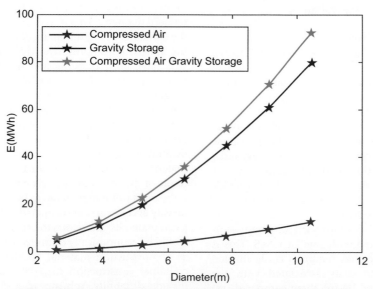

FIG. 2.20 Energy delivered as a variation of the container's diameter.

particularly its diameter. An inverse relationship exists between the container diameter and its withstood pressure; when one increases the other decreases. The power dependence between the container diameter and the pressure is one, while this dependence relation is two between the diameter and the compressed air volume. Therefore, increasing the diameter of the container rather than its withstood pressure would results in

higher energy capacity. Fig. 2.20 shows the variation of the storage capacity of CAGES according to a change in the container's diameter. It is assumed that the container allowable pressure does not change as a result of diameter variation.

More energy can be obtained with the inclusion of compressed air to larger GES systems used in large-scale application. Fig. 2.21 shows the capacity of energy

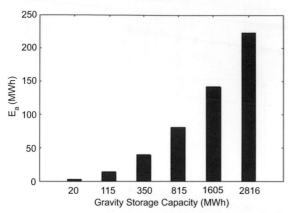

FIG. 2.21 Energy delivered from compressed air.

TABLE 2.4 Equipment Cost [40,42].	
Equipment	**Value in ($/Kw)**
Expander cost	220
Expander (low pressure)	140
Expander (high pressure)	60
Compressor cost	84
Heat exchanger cost	33
High pressure vessel cost	75
Hydroturbine cost	95
Pump cost	47

that might be released from the addition of compressed air to GES systems. For example, combining gravity storage with compressed air would lead to the addition of 223 MWh to a system of 2.8 GWh. Therefore, for large-scale GES systems, an attractive energy capacity could be added from compressed air.

System Characteristics

Compared with pumped hydro and CAES, CAGES offer a number of advantages, which make this new system an interesting one. These benefits include a simple structure offsetting the need of heating/cooling practices used by CAES. Furthermore, the anticipated CAGES efficiency is higher than that of CAES as it makes use of hydraulic equipment such as a hydroturbine. This latter is more efficient than the pressure expander used by CAES. In addition, hydraulic equipment are characterized by a rapid start-up, a capability to provide voltage and frequency regulations and an ability to perform spinning reserve services.

A lower capital cost of compressed air gravity storage is expected compared with that of CAES. This is because the system does not require a cooler/heater. Table 2.4 presents the costs associated with the different equipment used by the three compared systems. It is shown that the expander cost is much higher than the hydraulic mechanical systems. In addition to all these advantages, the installation of CAGES does not necessitate particular topographic conditions as required by PHS.

CONCLUSION

This chapter has presented a strategy to optimally design the different components of GES. The use of this methodology optimizes the storage capacity of

GES while taking into consideration the system design restrictions. To minimize the construction cost of this storage system, a NLP model has been presented in this chapter. The outcomes of this model demonstrate that for a particular storage capacity, it is more economic to design a container with a higher height and reduced diameter. That is, it is more practical to increase the container's height than its diameter while satisfying the system constraints. The different components of GES have been dimensioned in the presented parametric study. Optimal materials that should be used to construct these components were also discussed in this chapter. The identification of the piston optimal material was based on two important criteria, which are material density and cost. The system energy production is significantly affected by the density of the material used for the construction of the piston. The result of the material analysis has shown that iron ore is the most optimal material to be used due to its higher density and lower cost compared with the other investigated materials. Therefore, the energy generation of GES would be economically optimized by the use of iron ore as a construction material for the piston. As for the container construction material, the selection criteria include durability, robustness, and cost. To examine the stability of the system, an FEA analysis has been performed using SW. The results of this study demonstrate that steel is the best candidate to be used for the construction of container. However, for large-scale application, reinforced concrete is an alternative material that can be used because of its lower cost and long lifetime.

A methodology to identify the maximum valuable GES size has been proposed in this work. The aim of the sizing model is to maximize the profit by performing energy arbitrage service. This is done by identifying the optimum energy dispatch and operation of the energy storage system. This model takes into consideration the energy market variables, the hybrid plant constraint, as well as the technical and economic characteristics of the storage system. The case study outcomes validate the effectiveness of the optimization model. The maximum profitable storage capacity that should be installed by the owner has been determined using the proposed strategy. Therefore, the proper sizing of energy storage would enable wind farms to increase their profits while performing energy arbitrage with the electric grid. GES might be considered as an alternative to the current used energy storage system if it is optimally sized and designed with a goal to increase the system energy production and generated profit.

This chapter has also presented an improvement to GES system. The proposed system incorporates compressed air to GES. The energy that can be stored in this improved system has been determined using the proposed model. In addition, the impact of a number of parameters on the system performance has been investigated in this analysis. It has been shown that system storage capacity is influenced by the height and the diameter of the container. Furthermore, the energy delivered by CAGES can be improved by increasing the container withstand pressure.

Extra energy can be stored in CAGES as the air-to-water ratio is increased. A ratio of 1 would maximize the energy production of the system and hence results in an optimal configuration of CAGES. That is, compressed air should occupy half of the container volume. Interesting prospects regarding the storage capacity have been demonstrated by compressed air gravity storage. The presence of compressed air in GES would enable extra energy to be stored in the system.

NOMENCLATURE

A	Water conduit area (m^2)
A_A	Area of chamber A (m^2)
A_B	Area of chamber B (m^2)
A_c	Area of concrete section (m^2)
A_s	Area of circumferential reinforcement (m^2)
A_{st}	Area of steel (m^2)
C_c	Concrete cost ($)
C_F	Formwork cost ($)
C_{FU}	The formwork cost for a double face unit ($)
C_{cu}	Cost of concrete per unit $(€/m^3)$
C_S	Cost of steel per unit $(€/t)$
$C_{O\&M}$	Storage operation and maintenance cost in $(€/kWh)$
$Cost(t)$	Hourly costs of the hybrid farm

D	Diameter of the piston and inner diameter of the container (m)
Dp	Nominal outer diameter of pipe (mm)
e	Eccentricity
E	Energy production (J)
E_a	Compressed air released energy (MWh)
E_C	Excavation cost ($)
$E_D(t)$	Energy discharged from the storage at t (kW)
$E_G(t)$	Energy generated at time t
E_h	Hydraulic released energy (MWh)
$E_I(t)$	Energy sold/injected directly to the grid
E_L	Capacity limit of the storage system (kWh)
E_{UC}	Excavation unit cost ($)
E_V	Excavation volume (m^3)
$E_S(t)$	Energy stored at time t (W)
E_t	Total generated energy (MWh)
g	Gravitational acceleration (m/s^2)
H	High of the piston (m)
H_L	Height limit (m)
H_c	High of the container and return pipe (m)
H_t	Hoop tension
K	Cost variable $(€)$
k_c	Constant
K_0	Coefficient of earth pressure at rest
m	Modular ratio
m_r	Mass of the piston relative to the water (m/s^2)
P	Pressure (kPa)
P_1	Preset pressure (MPa)
P_2	Tank withstand pressure (MPa)
$P_E(t)$	Hourly energy prices $(€/kWh)$
Pr	Owner's profit which must be maximized
Px	Pressure of air in the container (MPa)
P_x^i	Internal pressure at location x (Pa)
P_x^e	External pressure at location x (Pa)
Q	Discharge rate
$Rev(t)$	Hourly revenues of the hybrid farm ($)
S	Ultimate strength of pipe material (kPa)
$S(t)$	Storage level (kWh) at time t (Wh)
$S(t-1)$	Storage remaining energy at time (t-1) (Wh)
$S_{Limit}(t)$	Storage capacity at time (Wh)
t	Thickness of the container
T	Nominal wall thickness of the pipe (mm)
Vx	Volume of air in the container (m^3)
V_1	Initial air volume (m^3)
V_2	Compressed air volume (m^3)
V_c	Concrete volume (m^3)
W	Specific weight of the water
xH	Specific location along the container height
y	Soil weight $(Kg/m3)$
z	Elevation height (m)
δ	Self-discharge rate of the system
η	Efficiency

μ' — Friction coefficient between the piston and the container's wall

ρ_p — Density of piston material (kg/m^3)

ρ_w — Density of water (kg/m^3)

σ_{cbc} — Compression stress developed in concrete (N/mm^2)

σ_{st} — Tensile stress developed in steel (N/mm^2)

β_c — Ratio of circumferential steel

β_v — Ratio of vertical steel

γ_S — Steel unit weight (t/m^3)

REFERENCES

[1] Libowitz GG, Whittingham MS. Materials science in energy technology. New York: Academic Press; 1979.

[2] Liu C, Li F, Ma L-P, Cheng H-M. Advanced materials for energy storage. Adv Mater 2010;22:E28−62. https://doi.org/10.1002/adma.200903328.

[3] Stanley Whittingham, M. Materials challenges facing electrical energy storage. USA: Binghamton University. Available from: https://www.mtixtl.com/documents/paper.pdf.

[4] Fernandez LA, Martínez M, Segarra M, Martorell I, Cabeza FL. Selection of materials with potential in sensible thermal energy storage. Sol Energy Mater Sol Cell 2010;94(10):1723−9.

[5] Deokar AJ, Yadav SD, Yadav SN. Material selection for biogas storage cylinder using fuzzy decision making method and FEA method. In: 2013 International conference on energy efficient technologies for Sustainability (ICEETS); 10−12 April 2013. p. 400−6. vol., no.

[6] Chen C, Duan S, Cai T, Liu B, Hu G. Optimal allocation and economic analysis of energy storage system in microgrids. IEEE Trans Power Electron 2011;26(10):2662. 773.

[7] Sharma S, Bhattacharjee S, Bhattacharya A. Grey wolf optimisation for optimal sizing of battery energy storage device to minimise operation cost of microgrid. IET Gener, Transm Distrib 2016;10(3):625−37. https://doi.org/10.1049/iet-gtd.2015.0429.

[8] Atwa M, El-Saadany E. Optimal allocation of ESS in distribution systems with a high penetration of wind energy. IEEE Trans Power Syst 2010;25(4):1815−22.

[9] Carpinelli G, Mottola F, Proto D, Russo A. Optimal allocation of dispersed generators, capacitors and distributed energy storage systems in distribution networks. In: Modern electric power systems conference; September 2010. p. 1−6.

[10] Changsong C, Duan S, Cai T, Liu B, Hu G. Optimal allocation and economic analysis of energy storage system in microgrids. IEEE Trans Power Electron 2011;26(10):2762−73.

[11] Pavković D, Hoić M, Deur J, Petrić J. Energy storage systems sizing study for a high-altitude wind energy application. Energy 2014;76:91−103.

[12] Zheng Y, Dong ZY, Luo FJ, Meng K, Qiu J, Wong KP. Optimal allocation of energy storage system for risk mitigation of DISCOs with high renewable penetrations. IEEE Trans Power Syst 2014;29(1):212−20.

[13] Musolino V, Pievatolo A, Tironi E. A statistical approach to electrical storage sizing with application to the recovery of braking energy. Energy 2011;36:6697−704.

[14] Feroldi D, Zumoffen D. Sizing methodology for hybrid systems based on multiple renewable power sources integrated to the energy management strategy. Int J Hydrog Energy 2014;39:8609−20.

[15] Schaede H, Schneider M, Rinderknecht S. Specification and assessment of electric energy storage systems based on generic storage load profile. In: 13 Symposium Energieinnovation; 2014. p. 1−10.

[16] Bennett CJ, Stewart RA, Lu JW. Development of a three-phase battery energy storage scheduling and operation system for low voltage distribution networks. Appl Energy 2015;146(15):122−34. https://doi.org/10.1016/j.apenergy.2015.02.012.

[17] Khorramdel H, Aghaei J, Khorramdel B, Siano P. Optimal battery sizing in microgrids using probabilistic unit commitment. IEEE Trans Ind Informatics 2016;12(2):834−43. https://doi.org/10.1109/TII.2015.2509424.

[18] Fossati JP, Galarza A, Martín-Villate A, Fontán L. A method for optimal sizing energy storage systems for microgrids. Renew Energy 2015;77:539−49.

[19] Zucker A, Hinchliffe T. Optimum sizing of PV-attached electricity storage according to power market signals−a case study for Germany and Italy. Appl Energy 2014;127:141−55.

[20] Pavković D, Hoić M, Deur J, Petrić J. Energy storage systems sizing study for a high-altitude wind energy application. Energy 2014;(76):91−103.

[21] Masih-Tehrani M, Ha'iri-Yazdi M-R, Esfahanian V, Safaei A. Optimum sizing and optimum energy management of a hybrid energy storage system for lithium battery life improvement. J Power Sources 2013;244:2−10.

[22] Beek AV. Advanced Engineering design. 2012.

[23] Civil engineering projects, see. http://www.civilprojectsonline.com/building-construction/reinforced-circular-water-tankdesign- of-rcc-structures/for Reinforced Circular Water Tank/Design of RCC Structures.

[24] Suresh GS. Design of water tanks. Mysore: National Institute Of Engineering.

[25] El Reedy MA. Construction management and design of industrial concrete and steel structures. Taylor & Francis Group; 2011. p. 576.

[26] Tareq M. Behaviour of reinforced concrete conical tanks under hydrostatic loading. University of Western Ontari; 2014. M.S. thesis.

[27] Indexmundi. Commodities market prices. 2015. Available at: http://www.Indexmundi.com/commodities/.

[28] Material cost. Available from: www.infomine.com.

[29] Cheremisinoff PN. Advances in environmental control technology. Houston, Texas: Storage Tanks Gulf Professional Publishing. 303.

[30] Gerard FJ. Gravity driven water network: theory and design. 2011.

[31] Berrada A, Loudiyi K. Operation, sizing, and economic evaluation of storage for solar and wind power plants. Renew Sustain Energy Rev 2016;59:1117—29.

[32] Nazari ME, Ardehali MM. Optimal coordination of renewable wind and pumped storage with thermal power generation for maximizing economic profit with considerations for environmental emission based on newly developed heuristic optimization algorithm. J Renew Sustain Energy 2016;8(065905). https://doi.org/10.1063/1.4971874 47.

[33] Abarghooee RA, Niknam T, Roosta A, Malekpour AR, Zare M. Probabilistic multiobjective wind-thermal economic emission dispatch based on point estimated method. Energy 2012;37:322.

[34] Aneke M, Wang M. Energy storage technologies and real life applications - a state of the art review. Appl Energy 2016;179:350—77.

[35] Ree. Red Eléctrica de España. S.A. (REE); 2017. http://www.ree.es/en.

[36] Chen H, Cong TN, Yang W, Tan C, Li Y, Ding Y. Progress in electrical energy storage system: a critical review. Prog Nat Sci 2009;19:291—312.

[37] Akinyele DO, Rayudu RK. Review of energy storage technologies for sustainable power networks. Sustain Energy Technol Assess 2014;8:74—91.

[38] Wang HR. Water-gas encompassing electric power, energy storage system: CN, 102434362A [P/OL]. 2012- 05-02 [2013-01-23].

[39] EBO Group, Inc. Compressed air pumped hydro energy storage and distribution system: US 7281371 B1 [P/OL]. 2007-10-16 [201-01-23].

[40] Wang HR, Wang LQ, Wang XB, Yao ER. A novel pumped hydro combined with compressed air energy storage system. Energies 2013;6:1554—1567.

[41] Bi J, Jiang T, Chen W, Ma X. Research on storage capacity of compressed air pumped hydro energy storage equipment. Energy Power Eng 2013;5:26—30.

[42] Drury E, Denholm P, Sioshansi R. The value of compressed air energy storage in energy and reserve markets. Energy 2011;36:4959. 497.

CHAPTER 3

Economic Evaluation and Risk Analysis of Gravity Energy Storage

INTRODUCTION

Balancing of energy demand and supply is properly done by the use of energy storage systems through the storing and discharging of energy. That is, the surplus of energy produced by renewable energy plants is stored by energy storage and dispatched when needed. In addition, energy storage systems provide multiple services to the electric grid such as regulation and grid stability. However, the development of these systems is still facing a number of issues such as their high capital cost. Although several research publications are available in literature discussing the economic aspects of different energy storage systems, a limited number of articles are about the economics of gravity energy storage (GES) [1—3]. An economic assessment of GES is covered in this chapter to investigate the viability of this system.

Multiple researchers have studied the economics of energy storage systems. It is challenging to compare these technologies on a common basis because of their different characteristics. The levelized cost of energy (LCOE) is a common approach used to compare the cost of energy storage system. As the storage system size increases, the system cost per kWh increases [4]. The levelized cost of storage (LCOS) for a 1 and 100 MW storage capacity of Li-ion batteries, pumped hydro storage (PHS), and compressed air energy storage (CAES) systems has been determined in Ref. [5]. The most cost-effective system in long-term application is PtG, while batteries are more appropriate for short time scale [5].

The cost per kWh to store energy for various energy storage systems is calculated in Ref. [6]. In a different study [7], the cost per charge/discharge cycle is compared for batteries, flywheels, PHS, capacitors, and CAES. LCOE of a photovoltaic (PV) system coupled to batteries including lead-acid, lithium ion, and redox-flow is dependent on the storage system C-rate [8]. A life-cycle cost analysis has been performed by Zakeri and Syri to investigate the cost of energy storage systems in different

grid applications, including arbitrage service, regulation, and transmission and distribution (T&D) support [9]. It has been shown by the authors that PHS and CAES have the least LCOE for energy arbitrage application. In addition, the most cost economic system for regulation service is flywheel energy storage. Vanadium redox flow battery (VRFB) LCOE has been calculated in Ref. [10]. The LCOS of adiabatic CAES and hydrogen storage system used in different applications including short-, medium-, and long-term has been determined in Ref. [11]. Hydrogen shows interesting prospects and could become cost-effective in 2030. A similar assessment has been performed for a 500-MW system [12]. The obtained results show that PHS has obtained a low LCOS of 2.5 €ct/kWh; followed by aCAES (5.3 €ct/kWh), and lead-acid batteries (15.9 €ct/kWh). A cost comparison of energy storage and fossil fuel alternatives has been conducted in Ref. [13]. The results of this study show that pumped hydro, CAES, and gas turbine have a completive LCOS in transmission investment deferral. In addition, the outcomes show that lithium-ions are considered economically viable in regulation application. Typically, the system capital expenditure (CAPEX) impacts significantly the storage LCOS. It is expected that some energy storage systems will become an alternative to natural gas power plants in the near term if the cost of energy storage continues to decline.

Only few literatures cover the economic aspects of GES system. That is why this chapter is devoted to the economic assessment of this novel storage technology. Section 2 presents a cost analysis of this system. This study identifies the different costs related to the construction of GES components, the acquisition of the hydro equipment, as well as operation and maintenance (O&M) costs. Next, a study to calculate GES LCOE is discussed. A comparison of energy storage LCOE is also presented. Section 3 identifies the value of energy storage in energy and ancillary service. A summary of the obtained results is presented in Section 4.

Gravity Energy Storage. https://doi.org/10.1016/B978-0-12-816717-5.00003-7

ECONOMIC ANALYSIS

The potential value of energy storage systems is considered a complex aspect, which is difficult to perceive because of the tailored nature of their economics. The performance of an economic study is necessary to investigate the viability of implementing energy storage systems. Considering only the preliminary value of installing, an energy storage system fails to take into account a number of factors that affect the total cost of these systems. The LCOE, also known as the LCOS, is a cost-oriented approach that accounts for different aspects such as the technology lifetime, efficiency, power, as well as O&M. This methodology is used also to compare the overall cost of energy storage with different characteristics. The LCOE of a storage technology is the ratio of the system lifetime cost and its life-cycle energy generation. The time value of money is considered by this methodology with a specific discount rate. To determine GES LCOE, the investment cost consisting of acquiring the mechanical equipment and constructing the system is considered. In this study, GES with a capacity of 20 MWh is used. The design and sizing of the system components have been performed based on this storage capacity. The container's height has been specified in this case as 500 m. In addition, it is assumed that the system discharge time is 4 h [14]. This later is used in the calculation of the flow rate, rated power, as well as the thickness and diameter of the return pipe. By taking into account the storage overall losses, which include hydraulic losses, frictional losses, and mechanical equipment inefficiencies, the storing of energy would take about 5h. The aforementioned data are used along with Eqs. (3.4, 3.8, 3.10, 3.11, and 3.12) presented in Chapter 2 to identify the system geometry as presented in Table 3.1.

The levelized cost of GES is calculated by summing the fixed and the variable cost of constructing, operating, and maintaining the system.

Cost of System Construction

The installation cost of gravity storage compromises the excavation and the construction costs of the system. The excavation cost includes both the cost of excavating the shaft and the return pipe; whereas the construction cost includes the cost of building the container structure, the piston, and the return pipe.

Cost of Excavation

To estimate the cost of excavating the system using a tunnel drilling machine, a literature review has been used due to the variation of this cost from one country to another. The tunnel cost in different countries has been identified by Hoek in Ref. [15]. This cost depends also on the type of the ground and soil conditions. The excavation cost ranges from 200 to 420 k€/m [16]. In this analysis, an average of 310 €/m^3 has been used for the calculation of the tunnel excavation unit cost. This cost is multiplied by the excavation volume of the shaft and the return pipe to obtain the overall excavation cost. The estimated excavation costs are equal to 1.99×10^7 € for the container and 83,100 € for the return pipe.

Material Cost for Piston

The material selection study discussed in Chapter 2 demonstrates that iron ore is the best candidate to construct the inner structure of the piston due to its high density and economic cost. To calculate the piston construction cost, the material cost per kilogram is multiplied to the piston mass. In this case study, the iron ore piston has a volume of 5347 m^3, which is equivalent to a mass of 42,084,813 kg. Therefore, the calculated cost has been found equal to 2 M€.

Material Costs for the Return Pipe and the Container

The construction material used in building the container structure has been investigated in Chapter 2. The criteria used in the identification of the most suitable material include toughness, durability, lifetime, and cost-effectiveness. These criteria are based on the fact that the structure should resist very high pressure caused by the piston and the water load, as well as the earth lateral pressure load. In this case, the most appropriate material that responds to all of the aforementioned conditions is reinforced concrete. The cost of construction consists of the material and labor costs. This has been neglected as it represents a small percentage of the total construction cost. Construction cost of the storage container has been discussed in Chapter 2 Section 2. The structure of container has been split into three parts, which include the structure wall, roof, and base. The container construction cost is a product of the concrete volume, the steel reinforcement, and the formwork. An estimation of the cost associated with the system construction is presented in Table 3.2.

TABLE 3.1
Parameters Used in Case Study.

Parameter	Height (m)	Diameter (m)	Thickness (m)
Container	500	5.21	7.6
Piston	250	5.21	—

TABLE 3.2
System Cost Estimation.

System Component	Cost (€)
Container	8.4×10^6
Return pipe	9.3×10^3

Equipment and Other Costs

The cost of acquiring the mechanical equipment of the powerhouse is also included in the storage system investment cost. This include the cost of the pump, motor, turbine, and generator. GES makes use of the same equipment used by PHS.

Hydroturbines are responsible for converting the kinetic energy of the flow to rotational energy. Various types of hydroturbines exist and are used in diverse applications. It is important to identify the most suitable type capable of working under specific conditions. The hydroturbine selection process should consider a number of aspects such as the system application, the flow variation, and the cost. The application criteria include the turbine head, characteristics, and rpm. The wide range of turbines that could be used in energy storage applications is restricted by the water head.

Reversible pump turbines are the most appropriate type for GES as they work in reverse direction. That is, this type of turbine can operate also as a pump. Francis and Kaplan reversible types could both be used in this storage application. However, reversible Francis turbine is more suitable for GES due to the benefits it offers compared with its counterpart. These include the system cost-effectiveness and better efficiency in both operating modes. The cost of equipment depends also on the installation mode, which includes centralized or decentralized type. The centralized unit makes use of a number of containers linked to one single powerhouse. This type of installation may result in pressure losses due the piping network it requires. In the decentralized type, each container is connected to its own powerhouse. Energy is generated separately from each energy storage. This latter type would be used in this case study. The cost of equipment including a 5-MW Francis reversible pump/turbine and a generator with a 50% contingency is 900 €/kW [17]. Therefore, the estimated equipment cost is about 4.5 M€.

Fig. 3.1 shows a cost division of GES investment cost. The highest share (57%) is represented by the excavation cost. This is followed by the construction and equipment costs with a share of 31% and 13%, respectively. In addition, a division of the construction

material cost including the container, the piston, and the return pipe costs are illustrated in Fig. 3.1. It appears that the container construction cost is higher than the other components. The construction cost of the return pipe is very low and constitutes only 1% of the system construction cost.

The mechanical equipment used by GES is similar to that of pumped hydroenergy storage, which includes reversible pump/turbine and a motor/generator. Therefore, the estimation of gravity storage balance cost and O&M costs is based on literature. The aforementioned costs have been valued as 1.9 €/kW for O&M cost and 4 €/kWh for system balance cost [17].

System Cost

The presented cost estimation takes into account only the direct cost of the system such as the raw material construction cost for building the piston and the container. The performed economic analysis does not consider the design, engineering, and installation costs of the technology. In addition, other components such as the sealing of the system have not been included as it is challenging to estimate the cost of a large sealing, which is not actually available in the market. This should be developed specifically for this technology. Consequently, this cost analysis is considered as preliminary; performing a comprehensive engineering analysis will accurately approximate the technology cost.

Gravity Storage Levelized Cost of Electricity

The LCOE calculation methodology discussed in Ref. [6] is used to determine the system LCOE. This is calculated by dividing the storage annual cost over the system expected energy production. An estimation of GES costs has been discussed before. The anticipated energy that would be discharged from the system is a function of the length and the number of charging/discharge cycles of the system per day.

The discharge length of energy storage depends on the grid applications. Storage systems used in generation application store/discharge energy for a period of 8 h per day as they typically charge energy at night and produce it during the day. The charging/discharging cycle for this application may typically be once per day. For the T&D application, storage systems usually discharge energy twice per day. The charging/discharging length is intended to be 4 h. It is assumed that energy storage used in this case study is used in T&D applications. This information along with the presented cost analysis would enable the calculation of GES LCOE used in T&D application. The LCOE has been found equal to 0.123 €/kWh.

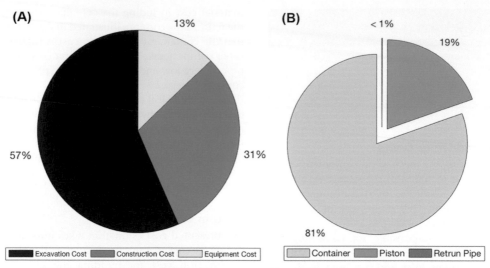

FIG. 3.1 Cost division of **(A)** system components and **(B)** structure construction cost.

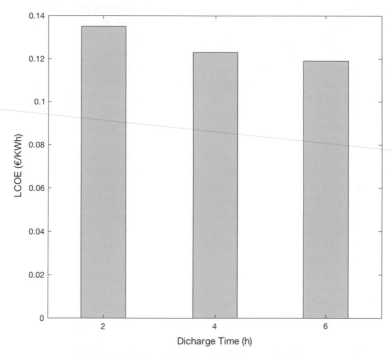

FIG. 3.2 System LCOE in generation applications.

Calculation of GES LCOE for different scenarios has been performed to evaluate the impact of the charging/discharging periods on the LCOE for both generation and T&D applications. A comparison of a number of scenarios for generation application is shown in Fig. 3.2. It can be seen that GES LCOE is significantly affected by the length of discharge. Likewise, Fig. 3.3 illustrates the LCOE for T&D applications with various charge/discharge periods. It is shown that as the discharging length increases, the LCOE decreases. It is important to mention that energy storage used in applications with higher number of cycles is more able to effectively participate in energy arbitrage services and hence increase their revenues. In addition, lower

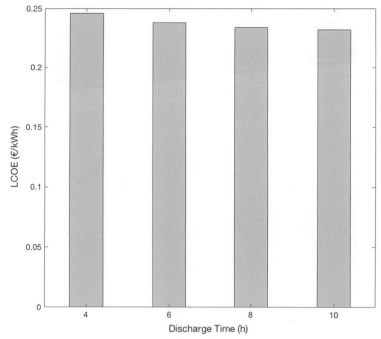

FIG. 3.3 System LCOE in generation applications in transmission and distribution applications.

cycle length enables these systems to benefit from peak energy prices and hence increase their profits. Consequently, energy storage has to participate in more grid services if used in applications with low discharge cycles to become economically viable.

Chapter 2 discussed three different sizing options for a GES system with a capacity of 20 MWh with a discharge time of 4 h. To determine the impact of system design and sizing on the LCOE, a comparison of these three-system LCOE is shown in Fig 3.4. It is shown that the LCOE is low of an optimally sized system. Therefore, the system design significantly affects the total system cost and LCOE.

Cost Comparison With Other Storage Systems

LCOE is considered a cost metric, which compares energy storage systems with different characteristics on a comparable basis. These systems have, for instance, unequal lifetime, capacities, rated power, capital cost, and efficiencies. The LCOE represents the lowest cost at which energy should be sold out so as to realize break-even over the storage life cycle. This approach allows for a fast and simple assessment of various energy storage systems.

LCOE quantification of different energy storage systems used in large-scale applications have been performed through a literature review [9–11]. The compared energy storage systems include lead-acid batteries, sodium-sulfur (NaS) batteries, underground and above CAES, nickel-cadmium (NiCd), iron-chromium (FeCr), PHS, and VRFB. The LCOE comparison of the aforementioned energy storage systems is shown in Fig. 3.5. In addition to the storage application and the discharge time, the LCOE is affected by a number of aspects such as the interest rate and technology characteristics.

The obtained outcomes indicate that PHS has the lowest LCOE (120 €/MWh), followed by GES with an LCOE of 123 €/MWh. Capital expenditure has the foremost noteworthy impact on LCOE delivered by pumped hydro, compressed air, and GES. The storage unit cost represents the highest share of capital cost for these three storage technologies.

A wide variety of LCOE is obtained for batteries due to the board variation in their capital costs. The highest LCOE, among all studied energy storage systems, is delivered by NiCd. The LCOE value of this latter system is about 421 €/MWh. This is followed by VRFs, which have a levelized cost of 353 €/kWh because of their high capital cost. Lead-acid batteries are ranked in the third place with an LCOE of 323 €/MWh. NaS and FeCr batteries have a lower LCOE compared with the other investigated batteries with an LCOE of 244 €/MWh and 209 €/MWh, respectively. The short lifetime of some batteries, such lead-acid, NiCd, and VRFB,

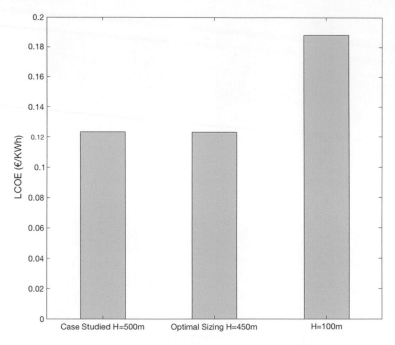

Different Sizing of Gravity Storage with E=20MWh

FIG. 3.4 System LCOE for different sizing options.

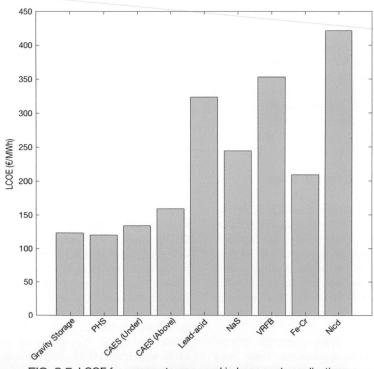

FIG. 3.5 LCOE for energy storage used in large-scale applications.

requires frequent replacement, which results in higher LCOE. Owing to the expected reduction in energy storage capital cost and anticipated system development, the LCOE of the aforementioned system is predicted to decrease in the next few years. A lower decrease is projected for mature technologies such as PHS and CAES. The LCOE of storage technologies that are at the market entry or still under development is expected to drop in the future.

The low LCOE obtained by pumped hydro and GES makes them the most cost-effective energy storage systems. Consequently, GES would be considered an attractive energy storage option due to its competitiveness with other energy storage systems.

The LCOE obtained for CAES and pumped hydro is very close to GES LCOE which makes these technologies comparable in bulk application. key performance features of each technology should be determined to obtain an overall comparison of these systems. Refs. [18−24] has been used to obtain these parameters. The compared characteristics include the investment and operation costs, the technology lifetime, as well as the LCOE are presented in Table 3.3.

The investment cost of an energy storage system is obtained by dividing the technology capital cost over its storage rated capacity. It is shown that GES has a higher investment cost compared to its counterparts. The operational cost which include the cost of O&M, repair, disposal, charging, and decommissioning. This cost is similar to all three investigated systems. It is estimated that the lifetime of GES is similar to that of PHS due to the number of similarities between these two systems. Typically, the life cycle of CAES is lower than that of PHS.

ENERGY STORAGE VALUATION IN DIFFERENT MARKETS

The challenge of storing electricity makes energy a real-time product, which has to be used as produced. To realize the simultaneity of energy consumption and generation, utility operators have to correctly balance and dispatch energy between generators and consumers. The balancing of energy in real time can be done through the use of energy storage. This latter system is able to store energy when there is a surplus and deliver it for consumption when needed. In addition, energy storage can provide energy to both ancillary and energy markets to make profit. They can participate in a number of applications by providing services such as regulation, arbitrage, and others [24]. Despite the benefits offered by energy storage system, their participation in a number of services is still not allowed by some markets, such as regulation and spinning/nonspinning reserves [25]. Because of the increasing integration of renewable energy systems coupled with energy storage, there is an interest in assessing the economic value of energy storage systems [26]. A number of energy storage technologies are not considered profitable in several applications and while providing some services [27]. Approaches to properly determine the benefits of installing energy storage systems are necessary to support the deployment of these systems.

The development of competitive energy markets has been followed by an increasing research devoted to energy storage value and economic profitability. The valuation of energy storage participating in one or more applications has been tackled by multiple research papers; however, it is still difficult to properly value some storage functionalities [28]. A value quantification of CAES in energy and reserve markets has been identified by Drury et al. in multiple US states [28]. The investment in this storage system is not supported in some markets by participating in only arbitrage service. However, the provision of reserve and arbitrage services enables conventional CAES to obtain a positive profit. Concerning the adiabatic type of CAES, this system is not profitable if it performs both energy arbitrage and reserve services.

The dispatch of energy has been modeled using various programming types [29]. The usual standard model makes use of linear programming (LP), whereas the powerful model uses mixed integer linear programming (MILP), especially for large-scale operation models [29]. Optimally operating a renewable farm combined with thermal energy storage using MILP results in an increase of revenues while performing day-ahead energy arbitrage and spinning reserve services [30]. A two-stage stochastic model is also used to optimally schedule the operation of wind farm coupled with PHS [31]. The aforementioned method is considered efficient for decision-making in real-time energy market under uncertainty. Sioshansi et al.

TABLE 3.3 Comparison of Energy Storage.			
System	GES	PHS	CAES
Cost (€/kW)	5840	600−5200	400−700
Operational cost (€/kW-yr)	1.9	1.9	1.9
Lifetime (yr)	40−60	40−60	20−40

examined the benefits that can be obtained in PJM market with an aim to identify the impact of storage capacity, system efficiency, fuel mix, and transmission constraints. It has been found that the value of energy storage is dependent on the market structure, the contract, and the ownership [32]. If potential value could be obtained by energy storage from participating in ancillary or capacity services, the colocation of energy storage and wind farm would become less attractive [33]. Further analysis is needed to determine the advantage of using CAES to perform ancillary services [33]. Authors in Refs. [34,35] have conducted an economic study to examine the value of combining wind plants with CAES and batteries, respectively. The outcomes of this analysis demonstrate that the combination of wind farm and batteries enables profit improvement.

A number of research analyses focusing on the value of energy storage have been performed in diverse locations. In PJM market, authors in Refs. [32−36] identified the potential benefits of PHS. Arbitrage revenues in European energy and North American markets have been studied in Ref. [37]. PJM and NYISO interconnections were examined by authors in Refs. [32,38], respectively. Regulation service and energy arbitrage have been valued in New York market by Walawalkar et al. Very attractive potential revenues can be obtained by energy storage performing regulation in NYISO. In addition, T&D system upgrade deferral service is an interesting opportunity for energy storage to generate additional revenues [38]. Multiple US wholesale markets were used in the study conducted to determine the value of energy storage [26]. The UK energy market was used by Locatelli et al. to determine large energy storage economics through an optimization model. This study reveals that subsidies are required for energy storage operating as price arbitrage and capacity reserve [39]. An economic analysis has been performed in the UK market by McKenna et al. to examine the benefits of incorporating lead-acid battery to PV system. The results revealed that the obtained profit is negative, and economic loss is important (around £1000/year) [40]. A self-scheduling methodology for energy storage systems has been proposed by Kazempour et al. to identify the maximum profit, which can be obtained in a number of markets [41]. In addition, a comparison between conventional and emerging systems has been presented by in this analysis. Authors in Refs. [42,43] have performed an economic study to determine the value of NaS and VRB energy storage systems in a competitive energy market. CAES valuation on the French energy market combines both regulated and deregulated revenue

opportunities [44]. The arbitrage value of this latter energy storage system is less compared with pumped hydrovalue [45]. However, in a different study conducted by Hessami and Bowly, CAES has been found to be the most economically profitable storage system compared with pumped seawater hydrostorage and thermal energy storage [46]. This study has been performed with an objective to identify the maximum obtained revenues from operating an energy storage system coupled with a wind farm in Poland. In ERCOT market, a study has been conducted by Fares and Webber to optimally operate an energy storage. This analysis has shown that the use of batteries to provide only wholesale energy arbitrage is unprofitable [47]. In addition, the cost of material degradation should always be considered while modeling the operation management of an energy storage system. The use of a price-maker energy storage in the electric grid has an impact on the market price, particularly during peak price hours [48].

Limiting the value quantification of energy storage to arbitrage service does not provide an accurate valuation of these systems. Interesting revenue opportunities can be obtained by energy storage from participating in ancillary service markets. This enables the real-time balancing of energy demand and supply, reliability, and grid stability. Such services include load following, regulation, and reserve capacity. One of the most valuable ancillary services for energy storage is regulation because of the fast response time of these systems [24]. Additional services that can be provided by energy storage include congestion relief, T&D upgrade deferral, backup capacity during outage. A comparison of renewable farm revenues with and without the participation of energy storage in ancillary markets has demonstrated that more revenues could be obtained if energy storage is allowed to perform ancillary services [6].

Optimal siting of energy storage in the grid has been modeled by Krishnan and Das to compare the benefits obtained from optimally allocating energy storage with the transmission expansion option. In the presented model, energy is dispatched to both energy and ancillary service markets [49]. Optimization of energy storage value for the performance in both energy and ancillary service markets has been performed by Das et al. CAES has been used in this study with the use of both unit commitment and economic dispatch programs. This latter program is operated after making commitment decisions [50].

In this chapter, the value of energy storage participating in several markets in day-ahead and real-time markets is examined to extend the available research in literature. The hourly operation of energy storage would be properly valued as it takes into account multiple revenue streams. A number of energy storage functionalities are provided in ancillary service; this study considers regulation. Reserve service has not been examined by the model because of its low occurrence compared with regulation Typically, the call for regulation service occurs around 400 times per day; although only 20 dispatches per year are set for spinning reserve [51].

Electricity Markets

The trading, selling, and purchase of energy occur in the retail and wholesale markets. Energy is purchased and sold between resellers and generators in the wholesale market. Resellers are also named as suppliers as they supply and sell energy to customers in the retail market. Resellers include electric utility companies, energy marketers, and competitive energy providers. Energy is purchased by resellers in the wholesale market form energy producers (generator) and sold in the retail market. The buying of energy is accomplished via

markets or contracts. These are established between sellers and costumers. These end users are offered a wide variety of options to choose from. They may purchase energy from a competitive supplier or from their local utilities. The energy market players are presented in Fig. 3.6. The purchase and the selling of power are indicated by the right-to-left ($) and the left-to-right arrows (MW and kW), respectively.

Energy Arbitrage

Energy demand is met by energy markets in real time. The energy market compromises both the real-time and the day-ahead markets. This is also known as forward market and is responsible for creating economic schedules 1 day before, for the generation and consumption of energy. The locational marginal prices (LMPs) for the next operating day are defined based on the generation offers, the demand bids, and the programmed bilateral transactions.

The real-time market is also known as the spot market; it is responsible for balancing energy quantities

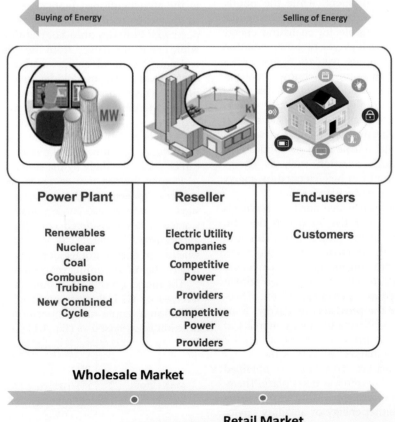

FIG. 3.6 Electricity markets players.

between the real-time requirement and the day-ahead scheduled energy. The LMPs of real-time market are determined based on the operating conditions of the actual electric grid. A number of potential opportunities are available in these markets for the provision of arbitrage service. This is due to the variation of energy prices enabling systems to buy energy during periods of low energy prices and sell it during peak energy demand when the prices go up. Energy consumption and production are counterbalanced by energy storage. The ability of storing and discharging energy has an important effect on the performance of the utility grid and the economic opportunities of energy market players.

Ancillary Services

Electric grid reliability and stability are ensured by grid operators through ancillary services markets. Ancillary services compromise a number of functionalities such as reserve capacity, regulation, and load following which can be performed by energy storage systems. Regulation services follow energy supply and demand fluctuations moment to moment, while the hourly fluctuations are met by reserve services. Therefore, the ancillary market is responsible for matching energy demand and supply, as well as assisting the operation of T&D system. Continuous regulation and quick response are provided by some generators characterized by automatic generation control. Provision of ancillary services is still prohibited for energy storage in some regions. Some energy storage devices such as flywheels contribute effectively to regulation services. Conversely, they should not be used in applications requiring long discharge duration such as energy arbitrage [36]. High profit can be generated by energy storage for the provision of ancillary services.

In this chapter, the investigated ancillary service is regulation that has an objective to match the grid voltage and frequency to the target value. This grid functionality controls the interconnection frequency, balance production with consumption, and matches the scheduled with the actual power flow. Requirement of regulation is defined as a percentage of the scheduled energy load. For the provision of energy, firms submit bids to both ancillary and energy markets. If the bid is successful, the firm receives a clearing price in the ancillary service market, as well as the energy price in the energy market. This latter is obtained only if the call for ancillary services takes place. Therefore, energy firms can opt for providing or promising to supply energy, to either energy or ancillary service markets.

Methodology

The objective of this study is to value energy storage while performing the most common grid applications, which include energy arbitrage and ancillary services. This quantification is done in multiple markets, which are: real-time and day-ahead energy markets, as well as regulation market. The value quantification of energy storage has been examined by a number of articles through the analysis of the system profitability while performing arbitrage in energy markets. However, such studies underestimate the value of these systems as they do not consider the possibility of energy storage to perform other grid applications such as ancillary services. The participation in these applications increases the system-obtained profit. This chapter discusses a methodology to value energy storage while taking into account several sources of revenues. The multiple considered sources come from three different markets.

This model has been developed to complement the currently available models in the literature, which take into consideration only one or two revenue opportunities. To investigate the profitability of energy storage in the aforementioned markets, a LP model is developed. Energy used to charge energy storage is either purchased from day-ahead or real-time energy markets; while energy discharged from the system is sold to the real-time/day-ahead energy markets, or the ancillary service market. The model identifies the optimal quantity of energy to store and discharge during each period. It also determines the capacity of energy to provide for regulation services. This optimal operation is based on the market and the storage state. The model provides the bidding decisions and the maximum profit, which can be generated daily by the storage system.

Despite the fact that energy storage advantages are properly understood, the valuation of those systems stays a controversial subject matter because energy storage is not considered cost-effective in some grid applications. Therefore, it is important to analyze whether an energy storage device has to perform some services before becoming economical.

The model objective function is to maximize project received by GES. Because profit is equal to the plant operation revenues minus operation cost, the objective function is expressed as (Eq. 3.1):

$$\text{MAX}\left[\text{Profit} = \sum_t \begin{bmatrix} P^{DA}(t)\left(E_d^{DA}(t) - E_c^{DA}(t)\right) + \\ P^{RT}(t)\left(E_d^{RT}(t) - E_c^{RT}(t)\right) + \\ P_{AS}^{RT}(t)E_{AS}^{RT}(t) + P^{RT}(t)\psi E_{AS}^{RT}(t) + \\ P_{AS}^{DA}(t)E_{AS}^{DA}(t) + P^{RT}(t)\psi E_{AS}^{DA}(t) - \\ C_d\left(E_c^{DA}(t) + E_c^{RT}(t)\right) - C_S \end{bmatrix}\right]$$

$$(3.1)$$

The storage system participates in energy arbitrage in both real-time and day-ahead markets, while performing regulation service. The first term in Eq. (3.1) represents revenues and costs in day-ahead and real-time energy markets, respectively. Revenues are received from the selling of energy, whereas costs represent the purchase of energy from the aforementioned energy markets. The second and third terms are about day-ahead and real-time ancillary service market revenues, respectively. Finally, the last term denotes the storage system costs.

Revenues obtained from participating in energy market are determined by multiplying the energy sold at a specific hour (h) with the price of energy at that time. On the other hand, revenues received from performing regulation service are determined differently. Two payments are provided for bidding in this ancillary market. These include a payment for reserve capacity and clearing price payment. The first one is paid irrespective to whether there has been a call for ancillary service or not, whereas the second one is obtained once the call for regulation service has been placed. This latter is provided for producing energy; it is a compensation per kWh of energy discharged. Ancillary service revenues in day-ahead market are formulated using Eq. (3.2):

$$R(t) = P_{AS}^{DA}(t)E_{AS}^{DA} + P_{AS}^{RT}(t)\psi E_{AS}^{DA}(t) \qquad (3.2)$$

Regulation service is monitored by utility grid operators. The occurrence for a regulation call is quite frequent per day and is about 400 times with different capacities [51]. The exchanged energy for this ancillary service is the portion of the actual (real-time) quantity of regulation and the capacity procured by the ISO [52]. This is known as the dispatch-to-contract ratio (ψ), which has been estimated by authors in Refs. [51,53] using historical data of a number of years. The average estimated ratio ψ is equal to 0.08–0.1. A ratio of 0.1 is used in this case study. This is equivalent to 0.1 kWh of energy being called in real-time regulation market, for each kilowatt of power traded in the regulation market for a period of 1 h. This ratio is assumed to be constant for the studied period. The aforementioned assumption may overstate the revenues obtained from providing regulation service as this ratio is variable in reality.

The value of energy storage is determined from the system operation revenues and costs. This also refers to the cost of providing energy and is calculated from the system capital expenses as well as degradation cost (C_d).

As explained in Chapter 2, the operation of energy storage is subject to the storage energy level constraint expressed in this case as (Eq. 3.3):

$$S(t) = S(t-1)(1-\delta) + \eta\left(E_c^{DA}(t) + E_c^{RT}(t)\right) - \left(E_d^{DA}(t) + E_d^{RT}(t) + \psi\left(E_{AS}^{DA}(t) + E_{AS}^{RT}(t)\right)\right) \qquad (3.3)$$

The storage state, $S(t)$, at a specific time, depends on the energy left in the storage system at (t-1), and the energy stored and discharged from the system at that time (t). System losses are taken into consideration by the model. These include the system round-trip efficiency and self-discharge rate.

The energy used to charge the storage system is purchased either from the real-time or day-ahead energy market. This energy is constrained by the system maximum charging/discharging rate, as well as the time fraction used to store/produce energy.

$$E_c^{DA}(t) + E_c^{RT}(t) \leq X_c(t)P_L \qquad (3.4)$$

$$E_d^{DA}(t) + E_d^{RT}(t) + \psi\left(E_{AS}^{DA}(t) + E_{AS}^{RT}(t)\right) \leq X_d(t)P_L \qquad (3.5)$$

The objective function of the model is to maximize profit by optimally operating energy storage. It has been explained in Chapter 2 that energy storage should not discharge and charge energy simultaneously at a given period. These functionalities should not be performed at the same time to avoid energy losses. This constraint has been expressed as:

$$X_c(t) + X_d(t) \leq 1 \qquad (3.6)$$

The storage energy level at time (t) should not exceed the system storage capacity and should be greater than the energy discharged at that specific time. This restriction is defined by (Eqs. 3.7–3.8).

$$S(t) \leq E_c \qquad (3.7)$$

$$E_d^{DA}(t) + E_d^{RT}(t) + \psi\left(E_{AS}^{DA}(t) + E_{AS}^{RT}(t)\right) \leq S(t) \qquad (3.8)$$

All energy vectors must be positive. The model is subject to this last construct, which is given by Eq. (3.9):

$$E_d^{DA}(t), E_d^{RT}(t), E_c^{DA}(t), E_c^{RT}(t), E_{AS}^{DA}(t), E_{AS}^{RT}(t), X_c(t), X_d(t) \geq 0 \qquad (3.9)$$

Case Study

To verify the effectiveness of the proposed valuation mode, a case study is presented in this section. General Algebraic Modeling System (GAMS) is used to develop and solve this LP mode. This case study makes use of NYISO market to obtain historical data [48]. NYISO is

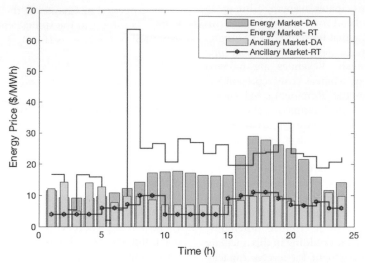

FIG. 3.7 Energy prices in all studied markets.

responsible for the administration of the wholesale power markets in New York. An optimization of the storage system operation is performed for 1 day (24 h) to assess the potential of performing energy arbitrage in real-time and day-ahead markets, as well as the potential opportunities in performing regulation service. GES with a capacity of 20 MWh is used in this case study. The degradation cost (C_d) is neglected because the system does not suffer from degradation as similar to PHS [54]. GES investment cost has been estimated as 5840 €/kW.

Day-Ahead, Real-Time, and Ancillary Energy Prices

Real-time and day-ahead energy prices for both energy and regulation markets are illustrated in Fig. 3.7. There is a wide difference between the price of energy in these markets. Table 3.4 shows that hourly energy prices in real-time market fluctuate more than in day-ahead energy market. It can be seen that these prices experience high variance and standard deviation. This volatility of prices in real-time energy market is due to the sudden issues that may arise in real time. Such events could be, for example, an unexpected change in weather, or an equipment outage in either generation or transmission station. It is to be noted that energy price be sometimes negative in real-time market. This is due to the fact that generators have to shut down during periods of low energy demand to avoid running at their low operating limit. This leads to an incurred unit startup charges.

Day-ahead prices may be higher than real-time prices because of two main reasons: safety and security. Some clients opt to pay more for electricity in day-ahead

TABLE 3.4
Market Statistical Characteristics of Energy Prices.

	Mean (Average)	Standard Deviation	Variance
Day-ahead energy market	16.63	6.056	36.68
Real-time energy market	22.19	11.51	132.5
Day-ahead ancillary market	9.045	2.401	5.769
Real-time ancillary market	6.62	2.55	6.505

energy market, so that they abstain dealing with excessive instability of prices in real-time energy market. Security is considered the second cause resulting in lower real-time prices. Some energy suppliers desire to keep a reserve energy capacity from day-ahead market and trade it in real-time market, so as to ensure security against the occurrence of outage. Conversely, day-ahead prices may become lower than real-time energy prices. This is because not all energy storage systems are allowed to participate in real-time energy and regulation markets. These markets necessitate a quick response time, which is not a characteristic of all energy systems. Consequently, these are prohibited in participating in these markets, which results in higher real-time energy prices.

An illustration of energy storage operation and participation in the studied markets is shown in Fig. 3.8. Energy is mostly dispatched by GES in the regulation market. Its participation in energy market both at day-ahead and real-time is limited and represents only a small share. Energy is mostly stored $(E_C^{DA}(t))$ in day-ahead energy market because of the low energy price it offers compared with real-time energy market. Table 3.4 demonstrates that energy prices fluctuate more in real-time market. This fluctuation leads to more opportunities to increase profit in real-time market. Providing more power to the regulation market is the most optimal operation strategy, which should be followed by energy storage. Energy storage participation in real-time and day-ahead energy markets occurs only when the storage level is low. The storage needs to be

charged particularly during periods when energy prices are low; as shown in Fig. 3.9. Energy is stored after being purchased from real-time or day-ahead market depending the price of energy. Therefore, to increase revenues, energy storage should deliver the majority of energy to the ancillary market.

Energy Storage Participation in Different Markets

GES participation in real-time and day-ahead energy markets is shown in Fig. 3.10A. Energy is bought during periods of low prices, particularly in the morning and early afternoon. The purchased energy in day-ahead energy market takes place between 7 a.m. and 4 p.m. This energy is sold out during late afternoon, evening, and morning; when the price of energy goes up.

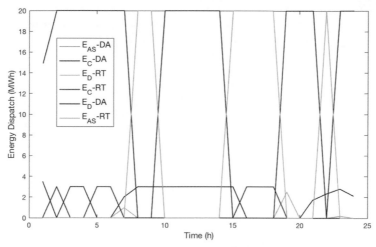

FIG. 3.8 Exchanged energy in different markets.

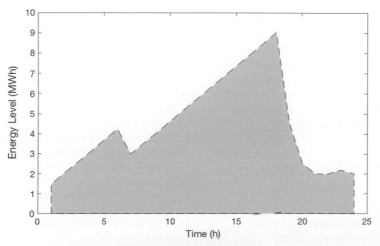

FIG. 3.9 Hourly storage state.

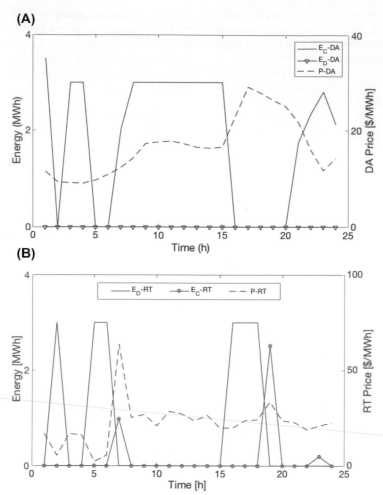

FIG. 3.10 Energy dispatch in **(A)** day-ahead market and **(B)** real-time market.

Energy is only traded to real-time energy market and regulation market. It is shown in Fig. 3.10 that energy is not sold to day-ahead market due to the low prices it offers compared with the other competitive markets. Energy discharged in real-time energy market occurs for a short time at 7 a.m., 7 p.m., and 11 p.m. (Fig. 3.10B).

Fig. 3.11 shows the participation of energy storage in ancillary service market both in real-time and day-ahead. Because of the better opportunities provided in day-ahead regulation market, more energy is offered to this market than in the real-time regulation market.

Storage Revenues

Revenues obtained from offering energy in both energy and ancillary markets are illustrated in Fig 3.12. The regulation market enables the storage system to make up to 97% of potential revenues; whereas the real-time energy market accounts for only 3%. No revenues are obtained in day-ahead market due to the absence of energy sold out in this market. Therefore, no profit comes from participating in day-ahead market (see Fig 3.13).

The two payments offered by the regulation market including the energy clearing price and the dispatching prices increase the potential revenues, which could be obtained from providing energy to this market. Conversely, the highest expenses are incurred by the day-ahead energy market. This is because of the fact that most of the purchased energy occurs in this market. Based on the discussed outcomes, a number of opportunities resulting in positive net present value arise from providing grid services such as regulation

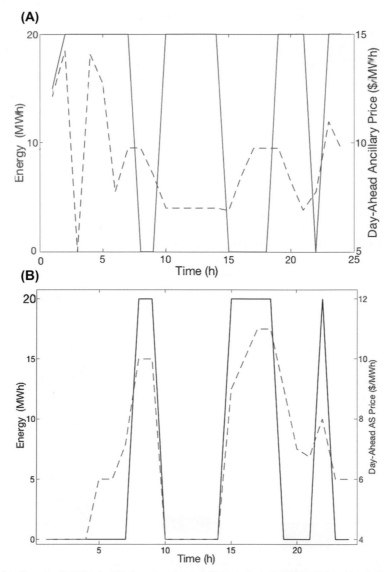

FIG. 3.11 Energy dispatch in **(A)** day-ahead regulation market and **(B)** real-time regulation market.

and arbitrage services; in both real-time and day-ahead markets. Therefore, various potential benefits can be achieved by the use of GES in the aforementioned applications. Typical, profit obtained from arbitrage and regulation relies on a number of factors including market flexibility and energy mix.

Currently, the permission to participate in some grid services, such the synchronous spinning reserve (10 min) is not given to a number of energy storage technologies. The aforementioned market provides 15% of revenue given by the regulation market [55].

Comparison to Competitive Energy Storage

The daily profit obtained from performing arbitrage and regulation services is shown in Fig. 3.14. It is shown that GES generates a negative profit due to the low revenues it receives from participating in these two services. The revenues are being exceeded by the system cost, which result in a negative NPV. This profit has been compared with the profit generated by alternative energy storage from performing the aforementioned services. These systems, which are mostly used in large-scale applications, include

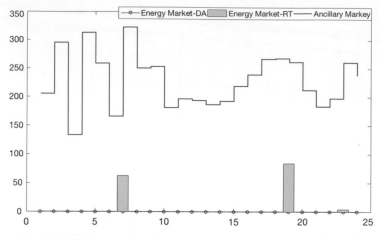

FIG. 3.12 Hourly revenues obtained in day-ahead and real-time markets.

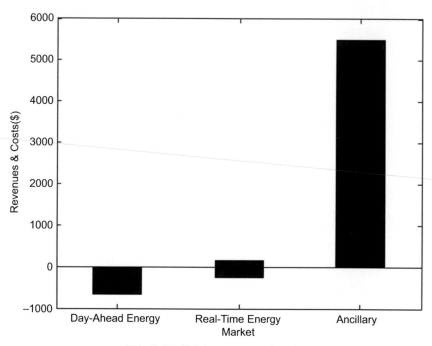

FIG. 3.13 Total revenues and costs.

pumped hydroenergy storage and CAES. Fig. 3.14 shows the comparison of the daily profit generated by these alternative storage systems. CAES has been able to make high profit compared with the other systems. In addition, a positive profit has been generated by PHS. This make these two technologies profitable while participating in energy and regulation markets. On the other hand, GES system is not considered economically profitable because of the negative NPV it generates.

Additional Benefits

A quantification of the various revenues provided to energy storage from participating in other grid services has been performed using literature review [56–58]. The value of power quality and reliability is typically done by commercial and industrial customers. Benefits received from end-user applications and regulated sectors have been quantified by Ref. [59]. Authors in Ref. [60] discussed the characteristics required for each specific application with its received benefits. Fig. 3.15

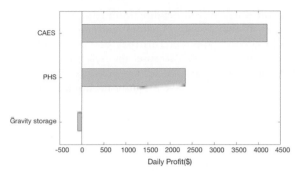

FIG. 3.14 Comparison of gravity storage with CAES and PHS.

shows the estimation of storage value in a number of applications based on literature review.

An important energy storage service that has not been investigated in this study is system upgrade cost deferral. This service enables grid utilities to delay the upgrade of T&D system by placing an energy storage system. This service may be performed for only some specific regions. The energy demand growth of these latter locations should be slow and variable, with high energy demand periods happening only for a short period of time during the day. System upgrade cost deferral value varies from 50 to 1000 ($/kW-year) depending on the utility market [57,58].

Although the involvement of energy storage in some grid applications enables the technologies to make some revenues, it is essential to perform a number of services to become economically profitable and support its use. To examine the profitability of GES from performing some applications, a cost-benefit analysis is performed. The results of this assessment are illustrated in Fig. 3.16. It is assumed, in this study, that GES is used for the specified applications including T&D upgrade deferral, generation or demand side capacity investment, and avoided distribution in outage. The obtained outcomes indicate that the system annual benefits received from performing the three aforementioned services are being exceeded by the storage technology annual cost. Therefore, GES is not economically viable if its deployment is limited to these services. The system value was underestimated in this case. Accordingly, it is important to capture the full benefit of deploying an energy storage system by allowing it to participate in a number of applications. A discussion about various grid applications will be presented in Chapter 4.

RISK ANALYSIS

Investment in sustainable energy projects might be viewed as unsafe and risky because of the financial, economic, and political conditions of a country. The viability of the investment may be affected by additional risks such as the technology itself. Therefore, a decrease in investment interest is expected because of the project risks [61]. It is crucial to examine the different aspects that may affect the project viability and recognize the extent to which these factors could impact the project investment.

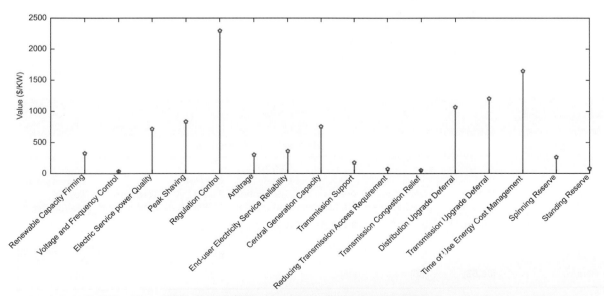

FIG. 3.15 Energy storage value in diverse applications.

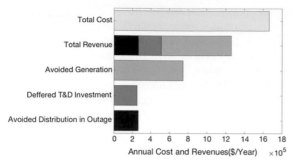

FIG. 3.16 Annual storage benefits versus costs.

In this chapter, a literature review has been used to identify GES investment risks [55,62–66] because of the similarity of this system to pumped hydroenergy storage. Risks associated with GES project investment are grouped into sets that include internal and external risks. Risks related to the performance of the system are classified as internal risks, whereas external risks are about the external aspects of the projects.

Internal/Technological Risks

Malfunction of the system is considered an internal risk. For GES system, risks may be mainly associated with a failure in the construction of the system or a fault in the mechanical equipment used. As for the system components construction, it is difficult to deeply excavate and place the system underground. Large-scale systems may reach a depth of 2000 m underground, which may be challenging to achieve. However, the performance of a suitable geotechnical analysis would eliminate this risk and reduce the construction failure. The system sealing is another issue as the system uses a large sealing, which is not common. In addition, buckling inside the container may occur in case the piston gets stuck inside.

This will result in wall bending moments. Breach and leakage in the system pipelines and the container should also be expected. This may be due to fatigue failures or corrosion. Because the system operates with high pressure and makes use of fluids, the aforementioned risks may have a significant impact on the system, leading to malfunction [67]. In general, risks may result because of components not well constructed or designed.

Mechanical equipment malfunctioning would also lead to technology operation failure. GES makes use of similar equipment used by PHS, which include a reversible pump turbine, a motor-generator, and transformers. A fault in one of these equipment may cause the system to stop operating. This will result in a loss of revenues due to the nonoperation of the technology for a specific period (time necessary to repair the fault).

Additional risks may result in the control of these equipment. Faults may occur in wires connecting the different equipment, in the control house, or in sensors. This latter role is to provide data about the system temperature, pressure, and others. An error in one of these parameters could lead to wrong decisions in the system operation. The use of reliable and good quality sensors is recommended in this case. The control room may encounter also some issues such as security and fire. Finally, the electric grid connection is another risk, which would impact the system operation. Example of issues include the malfunction of the transmission wires.

A quantification of the internal risk occurrence has been performed to rank them according to their occurrence per year. Tables 3.5 and 3.6 present the classification of these risks with their occurrence based on literature review [66,68,69].

This evaluation reveals that the system possibility of failure is fairly small. The system components with a high risk of malfunction include the transformer,

TABLE 3.5
Occurrence per Year of Internal Risks.

	EQUIPMENT MALFUNCTION						
Risk	Pump/ Turbine	Generator Malfunctions	Sensor Malfunction	Control Panel Connection Malfunction	Control Room Malfunction	Transformer Malfunction	Transmission Tower Malfunction
Occurrence per year	2.4×10^{-3}	3×10^{-4}	3×10^{-5}	8×10^{-4}	2×10^{-4}	3×10^{-3}	7×10^{-4}
Rank	2	5	8	3	6	1	4
Refs	[66,68]	[66]	[68]	[66,68]	[68]	[68]	[68]

TABLE 3.6
Occurrence per Year of Internal risks.

Risk	STORAGE COMPONENTS FAILURE		
	Pipeline Leak	Breach of the Pipelines (per Meter)	Breach of Shaft due to Corrosion or Fatigue (per Meter)
Occurrence per year	6×10^{-5}	3×10^{-7}	1×10^{-7}
Rank	7	9	10
Refs	[68]	[66,69]	[66,69]

pump/turbine, and control panel connection. To decrease the occurrence of these risks, regular maintenance and standard verification of equipment are required. In general, based on the obtained results, the aforementioned risks can be controlled easily, and the occurrence of serious operation issues is rather low.

External Risks

The project development and system operation are also subject to external risks. The most critical ones include political, economic, completion, financial, geologic, technological, sociocultural, force majeure, and environmental risks.

Financial risk may have an impact on the project financial condition. This type of risk is expected to occur during the realization phase of the project [70]. This risk is about interest rate uncertainty, inflation, and difficulties in procuring the project financing. A decrease in project investment is typically caused by a reduction in loans. In addition, variation in exchange rate also affects the financial situation of the project. A number of economic measures are able to control the inflation and the interest rate risk. The probability of an increase in these is low, and the likelihood of their detection is high. The impact of interest rate variability on the project profitability is investigated in the chapter 4.

Political risks could have also an effect on the project profitability. Examples of this risk include a change in existing legislation or government policies. These could have an impact on the project investment and lead to uncertainty among investors. In the energy sector, energy price volatility is considered as a market risk. This occurs due to the increase of renewable power plants. Trading in energy market is influenced by extreme energy price movements; this has also an

impact on project investment decisions [71]. Energy price volatility also affects the project profitability as energy markets are considered one of the revenue sources of renewable energy plants; the probability of energy price instability is moderate to high [72] and its impact is rather low [73].

Not completing the construction of the project while respecting the deadline is considered an important risk which should be considered in the project risk analysis. This risk study should be performed with respect to the project specified budget and specifications [72]. A number of aspects could result in uncompleted project such as design errors, cost underestimation, suppliers' issues, contractual problems, and unforeseen faults and failures in the different phases of the project. The operation of the project would be postponed because of the construction delay resulting in a loss of revenues. This will affect the project profitability. The occurrence and severity of the completion risk is moderate; while its detection is rather low.

An economic risk may be present in a project because of a number of factors; these include project mismanagement, cost increase, or a change in some economic aspects of the project. This risk may impact the project viability as it will make it challenging for investors to recuperate their input money throughout the system life cycle [74]. The occurrence of this risk is low, and its detection is not clear.

As GES requires deep underground excavation, geologic risks may be expected for this storage system. The implementation site of this GES should have the ability to accommodate the excavated system. A proper geologic study has to be conducted to determine the condition of the site, and any issues that may occur beforehand. Such issue includes seismic activities or site flows, which could impact the investment cost of the system. Furthermore, it is important to consider environmental risks linked with the development of GES. Oppositions associated with the environment may cause a delay in the construction of the system. The likelihood of occurrence and detection of this risk is low.

Additional risks that are about issues occurring as a result of unexpected events such as strikes, fires, catastrophic, and surges are classified as force majeure risk. This type of risk can impact the operation of the storage system. Its occurrence is typically low to moderate, and its detection rate is high [72].

Promotors and investors' interest is reduced with sociocultural risk. This risk is about the social and cultural diversity and differences between people in charge of the project, authorities, and workers. Issues resulting

TABLE 3.7
Rating Scale.

Rating	1–2	3–4	5–6	7–8	9–10
Occurrence	Minor	Low	Moderate	High	Extreme
Detection	Very high	High	Moderate	Low	Very low
Severity	Minor	Low	Moderate	High	Extreme

from this risk have an important impact on the project profitability as they may contribute to the nonoperation of the project due to, for example, boycotts and complaints. The probability of sociocultural risk is low while its severity is high. Other severe consequences of this risk include abandonment of project abandonment, revenue reduction, and investors reputational damage.

A classification of risks that have an impact on the economic viability of energy storage systems has been performed by Locatelli et al. [73]. The storage systems investigated in this study have a number of similarities with GES. One of the most significant risks that affects the system profitability is the unpredictability and the fluctuation of energy prices. In addition, the delay in the construction of the system along with the cost overrun have high impact on the system NPV. The probability of occurrence of these risks is high because of excessive capital cost incurred by GES similar to PHS [73].

A failure mode and effects analysis is performed in this section to determine which risks are more critical than others. A risk priority number (RPN) is calculated in this study by multiplying the occurrence, detection, and severity of the risk [72]. The occurrence rating refers to likelihood of the risk occurrence. The detection rating is about the capability of detecting the risk before its occurrence. The consequences resulting from the risk is measured by the severity rating [75,76]. A high RPN indicates a high risk and a high probability of system failure. Risks with high RPN should be addressed to reduce the chance of failure. Table 3.7 presents the scale used in the rating of the discussed risks.

External risks are ranked in Tables 3.8 and 3.9 according to the calculated RPN. Special consideration should be provided to risks with an RPN above 125 [72]. In this case, it is important to address these risks to create satisfactory circumstances for the investment in such project. These high risks are completion, economic, and political risks. Low-risk factors include force majeure risks, sociocultural, and environmental risks.

TABLE 3.8
External Risks Ranking.

	FINANCIAL RISK		
Risk	Interest Rate	Inflation	Exchange Rate
Occurrence	4	4	4
Detection	4	4	4
Severity	6	6	6
RPN	96	96	96
Rank	4		
Refs	[66,72]		

CONCLUSION

GES LCOE was determined in this chapter by the performance of an economic study. The LCOE enables the comparison of different energy storage systems. The LCOE approach was used to explore whether or not GES is a competitive system. The LCOE is determined by calculating GES lifetime costs, which include construction cost, M&O costs divided by the system's lifetime energy production. An estimation of these costs has been done to calculate the system LCOE in large-scale application. The LCOE methodology considers the time value of money along with a particular interest rate. Furthermore, a comparison of the obtained LCOE with that of other energy storage systems has been conducted. The outcomes of this study demonstrate that GES has an interesting LCOE compared with other systems; it is more or less similar to that of PHS. Therefore, this novel technology may be considered as a substitute to PHS in the near term.

To determine energy storage profit, an operation model has been discussed in this chapter. Energy storage revenues can be obtained from performing energy arbitrage in both real-time and day-ahead markets, in addition to ancillary services. The proposed

TABLE 3.9
External Risks Ranking.

Risk	Political Risk	Completion Risk	Economic Risk	Geologic Risk	Environmental Risk	Sociocultural Risk	Force Majeure Risk
Occurrence	5	6	7	3	4	3	4
Detection	4	4	8	4	3	2	8
Severity	9	6	7	3	4	8	2
RPN	180	144	392	36	48	48	64
Rank	2	3	1	7	6	6	5
Refs	[72]	[72,73]	[72,73]	[66]	[72]	[72]	[66,72]

model identifies the optimal energy capacity, which should be provided to the regulation service; along with the optimal energy quantity, which should be traded in the day-ahead and the real-time energy markets. The obtained model outputs reveal that performing regulation service enables energy storage to generate a high revenue. However, the total obtained revenues from the three investigated markets are still low compared with the system cost, which results in a negative profit. A comparison of GES NPV with that of PHS and CAES has shown that GES is considered less appealing in these applications.

Energy storage is able to capture additional benefits from participating in other grid services. According to the presented economic analysis, GES is not economically viable if it performs a limited set of services. Consequently, although the use of GES in a renewable energy plant increases its revenues, the storage system has to participate in multiple grid application to become profitable and justify its development.

NOMENCLATURE

C_d Energy storage degradation cost ($/MWh)
C_S Energy storage costs ($)
d Energy storage self-discharge rate
DoD   Depth of discharge
$E_{AS}^{DA}(t)$ Energy offered to regulation service in day-ahead ancillary market (MWh)
$E_{AS}^{RT}(t)$ Energy offered to regulation service in real-time ancillary market (MWh)
$E_c^{DA}(t)$ Energy purchased at time t in day-ahead energy market (MWh)
$E_c^{RT}(t)$ Energy purchased at time t in real-time energy market (MWh)

$E_d^{DA}(t)$ Energy sold at time t in day-ahead energy market (MWh)
$E_d^{RT}(t)$ Energy sold at time t in real-time energy market (MWh)
E_c Energy storage capacity (MWh)
L_C Energy storage lifetime in cycles
LET   Lifetime energy throughput (MWh)
$P^{DA}(t)$ Hourly energy price in day-ahead energy market ($/MWh)
$P^{RT}(t)$ Hourly energy price in real-time energy market ($/MWh)
$P_{AS}^{DA}(t)$ Hourly energy price in day-ahead ancillary market ($/MWh)
$P_{AS}^{RT}(t)$ Hourly energy price in real-time ancillary market ($/MWh)
P_L Energy storage power rating (MW)
$R(t)$ Hourly revenues ($)
$S(t)$ Storage level at time t (MWh)
$X_c(t)$ Charging period at time t
$X_d(t)$ Discharging period at time t
δ Storage self-discharge
ψ Average dispatch to contract ratio (MWh/MW)
η Energy storage round-trip efficiency.

REFERENCES

[1] Oldenmenger WA. Highrise energy storage core: feasibility study for a hydro-electrical pumped energy storage system in a tall building (Master's thesis). Retrieved from. TU Delft Repositories; 2013.
[2] Berrada A, Loudiyi K, Zorkani I. Profitability, risk, and financial modeling of energy storage in residential and large scale applications. Energy 2017;119:94–109.

[3] Tarigheh A. Master thesis: gravity power module (Master Thesis). Delft: Delft University of Technology; 2014.

[4] Härtel P, Doering M, Jentsch M, Pape C, Burges K, Kuwahata R. Cost assessment of storage options in a region with a high share of network congestions. J. Energy Storage 2016;8. https://doi.org/10.1016/j.est.2016.05.010.

[5] Abdon A, Zhang X, Parra D, Patel MK, Bauer C, Worlitschek J. Techno economic and environmental assessment of energy storage technologies for different storage time scales. In: International renewable energy storage conference IRES; 2016.

[6] Berrada A, Loudiyi K. Operation, sizing, and economic evaluation of storage for solar and wind power plants. Renew Sustain Energy Rev 2016;59:1117—29.

[7] Ibrahim H, Ilinca A, Perron J. Energy storage systems dcharacteristics and comparisons. Renew Sustain Energy Rev 2008;12(5):1221—50.

[8] Pawel I. The cost of storage e how to calculate the levelized cost of stored energy (LCOE) and applications to renewable energy generation. Energy Procedia 2014;46: 68—77.

[9] Zakeri B, Syri S. Electrical energy storage systems: a comparative life cycle cost analysis. Renew Sustain Energy Rev 2015;42:569—96.

[10] Viswanathan V, Crawford A, Stephenson D, Kim S, Wang W, Li B, et al. Cost and performance model for redox flow batteries. J Power Sources 2014;247:1040—51.

[11] Fichtner. Erstellung eines Entwicklungskonzeptes Energiespeicher in Niedersachsen. 2014 [Stuttgart].

[12] Weiss T, Meyer J, Plenz M, Schulz D. Dynamische Berechnung der Stromgestehungskosten von Energiespeichern für die Energiesystemmodellierung und-einsatzplanung. Z Energiewirtschaft 2016;40(1):1—14.

[13] Lazard. Lazard's levelized cost of storage analysis - version 1.0. 2015.

[14] Gravity power. Gravity power module. Energy storage. Grid-scale energy storage. 2011. Available at: http://www.gravitypower.net/.

[15] Hoek E. Big tunnels in bad rock. ASCE J. Geotechnical Geoenviron. Eng. 2001;127(9):726—40.

[16] Madlener R, Specht J. An exploratory economic analysis of underground pumped-storage hydro power plants in abandoned coal mines. Institute for Future Energy Consumer Needs and Behavior (FCN); February 2013.

[17] ALMÉN J, FALK J. Subsea pumped hydro storage- a technology assessment (Master's thesis). Sweden: Chalmers University of Technology; 2013.

[18] Akinyele D, Rayudu R. Review of energy storage technologies for sustainable power networks. Sustainable Energy Technol. Assess. 2014;8:74—91.

[19] IRENA. Electricity storage—technology brief. IEA-ETSAP and IRENA; 2012.

[20] Rastler D. Electricity energy storage technology options. Palo Alto: Electric Power Research Institute (EPRI); 2010.

[21] Evans A, Strezov V, Evans TJ. Assessment of utility energy storage options for increased renewable energy penetration. Renew Sustain Energy Rev 2012;16:4141—7.

[22] Independent Statistics & Analysis U.S. Energy Information Administration, see http://eia.gov [for Electricity Storage Cost].

[23] IRENA, for electricity storage—technology brief. 2012. see, https://www.irena.org/DocumentDownloads/Publications/IRENAETSAP%20Tech%20Brief%20E18%20Electricity-Storage.pdf.

[24] Denholm P, Ela E, Kirby B, Milligan M. The role of energy storage with renewable electricity generation. Golden, CO: National Renewable Energy Laboratory; 2010.

[25] House Economic Matters Committee, Maryland General Assembly. Energy storage: considerations for Maryland. January 2016. Available at: http://energy.maryland.gov/Reports/FY15_Energy_Storage_Report.pdf.

[26] Bradbury K, Pratson L, Patiño-Echeverri D. Economic viability of energy storage systems based on price arbitrage potential in real-time U.S. electricity markets. Appl Energy 2014;114:512—519. https://doi.org/10.1016/j.apenergy.2013.10.010 [ISSN 03062619]. http://linkinghub.elsevier.com/retrieve/pii/S0306261913008301.

[27] Schubert, E., Zhou, S., Grasso, T., Niu, G., A primer on wholesale market design. Market Oversight Division White Paper. Available at: https://www.hks.harvard.edu/hepg/Papers/TXPUC_wholesale.market.primer_11-1-02.pdf.

[28] Drury E, Denholm P, Sioshansi R. The value of compressed air energy storage in energy and reserve markets. Energy 2011;36(2011):4959—73.

[29] Ommen T, Markussen WB, Elmegaard B. Comparison of linear, mixed integer and non-linear programming methods in energy system dispatch modelling. Energy 2014. https://doi.org/10.1016/j.energy.2014.04.023.

[30] Pousinho HMI, Silva H, Mendes VMF, Collares M, Cabrita PC. Self-scheduling for energy and spinning reserve of wind/CSP plants by a MILP approach. Energy 2014;78:524—34.

[31] Garcia-Gonzalez J, R. de la Muela RM, Santos LM, Gonzalez AM. Stochastic joint optimization of wind generation and pumped-storage units in an electricity market. IEEE Trans Power Syst 2008;23(2):460—8.

[32] Sioshansi R, Denholm P, Thomas J, Weiss J. Estimating the value of electricity storage in PJM: arbitrage and some welfare effects. Energy Econ 2009;31(2):269—77.

[33] Denholm P, Sioshansi R. The value of compressed air energy storage with wind in transmission-constrained electric power systems. Energy Policy 2009;37(8): 3149—58.

[34] DeCarolis JF, Keith DW. The economics of large-scale wind power in a carbon constrained world. Energy Policy 2006;34(4):395—410.

[35] Bathurst GN, Strbac G. Value of combining energy storage and wind in shortterm energy and balancing markets. Electr Power Syst Res 2003;67(1):1—8.

[36] Perekhodtsev D. Two essays on problems of deregulated electricity markets. Ph.D. Dissertation. Pittsburgh, PA: Tepper School of Business Carnegie Mellon University; 2004. 99pp. Available at: http://wpweb2.tepper.cmu.edu/ceic/theses/Dmitri_Perekhodtsev_PhD_Thesis_2004.pdf.

[37] Figueiredo C, Flynn P, Cabral E. The economics of energy storage. In: Proceedings of the 2005 annual meeting of energy storage association, May 23—26, 2005, Toronto, Canada; 2005.

[38] Walawalkar R, Apt J, Mancini R. Economics of electric energy storage for energy arbitrage and regulation in New York. Energy Policy 2007;35 (1):2558—2568. https://doi.org/10.1016/j.enpol.2006.09.005 [ISSN 03014215], http://linkinghub.elsevier.com/retrieve/pii/S0301421506003545.

[39] Locatelli G, Palerma E, Mancini M. Assessing the economics of large Energy Storage Plants with an optimisation methodology. Energy 2015;83:15—28.

[40] McKenna E, McManus M, Cooper S, Thomson M. Economic and environmental impact of lead—acid batteries in grid-connected domestic PV systems. Appl Energy 2013;104:239—49.

[41] Kazmpour SJ, Moghaddam MP, Haghifam MR, Yousefi GR. Electric energy storage systems in a market-based economy: comparison of emerging and traditional technologies. Renew Energy 2009;34(12):2630—9.

[42] Kazmpour JS, Moghaddam MP. Economic viability of NaS battery plant in a competitive electricity market. In: 2009 International conference on clean electrical power; 2009. p. 453—9.

[43] Mogaddam IG, Saeidian A. Self scheduling program for a VRB energy storage in a competitive electricity market. In: 2010 International conference on power system technology (POWERCON); 2010. p. 1—6.

[44] He X, Lecomte R, Nekrassov A, Delarue E, Mercier E. Compressed air energy storage multi-stream value assessment on the French energy market. In: PowerTech, 2011. Trondheim: IEEE; 2011. p. 1—6.

[45] Siohansi R, Denholm P, Jenkin T. A comparative analysis of the value of pure and hybrid electricity storage. Energy Econ 2011;33(1):56—66.

[46] Hessami M-A, Bowly DR. Economic feasibility and optimisation of an energy storage system for Portland Wind Farm (Victoria, Australia). Appl Energy 2011;88(8):2755—63.

[47] Fares R, Webber M. A flexible model for economic operational management of grid battery energy storage. Energy 2014;78:768—76.

[48] Shafiee S, Zamani- Dehkordi P, Zareipour H, Knight MA. Economic assessment of a price-maker energy storage facility in the Alberta electricity market. Energy 2016;111:537—47.

[49] Krishnan V, Das T. Optimal allocation of energy storage in a co-optimized electricity market: benefits assessment and deriving indicators for economic storage ventures. Energy 2015;81:175—88.

[50] Das T, Krishnan V, McCalley JD. Assessing the benefits and economics of bulk energy storage technologies in the power grid. Appl Energy 2015;139:104—118. https://doi.org/10.1016/j.apenergy.2014.11.017 [ISSN 03062619]. http://linkinghub.elsevier.com/retrieve/pii/S0306261914011660.

[51] Kempton W, Tomic J. Vehicle-to-grid power fundamentals: calculating capacity and net revenue. J Power Sources;144:268-270.

[52] Xi X, Sioshansi R. A dynamic programming model of energy storage and transformer deployments to relieve distribution constraints. Comput Manag Sci 2016;13(1):119—46.

[53] Tomic J, Kempton W. Using fleets of electric-drive vehicles for grid support. J Power Sources 2007;168:459—68.

[54] Ecofys. Energy Storage opportunities and challenges. 2014. A West Coast perspetive White paper. Available at: http://www.ecofys.com/files/files/ecofys-2014-energy-storage-white-paper.pdf.

[55] Carneiro P, Ferreira P. The economic, environmental and strategic value of biomass. Renew Energy 2012;44:17—22. https://doi.org/10.1016/j.renene.2011.12.020.

[56] Butler P, Iannucci J, Eyer J. Innovative business cases for energy storage in a restructured electricity marketplace. 2003. Sandia National Laboratories report SAND2003-0362. Available at: http://www.prod.sandia.gov/cgi-bin/techlib/access-control.pl/2003/030362.pdf.

[57] EPRI. EPRI-DOE handbook of energy storage for transmission and distribution applications. Palo Alto, CA: EPRI; 2003. and the US Department of Energy, Washington DC.

[58] Eyer J, Iannucci J, Corey G. Energy storage benefits and market analysis handbook: a study for the DOE energy storage systemsprogram. Sandia National Laboratories; 2004. SAND2004- 6177. Available at: http://www.prod.sandia.gov/cgi-bin/techlib/access-control.pl/2004/046177.pdf.

[59] SANDIA. Energy storage for the electricity grid: benefits and market potential assessment guide. Albuquerque, New Mexico: Sandia National Laboratories; 2010.

[60] Beaudin M, Zareipour H, Schellenberglabe A, Rosehart W. Energy storage for mitigating the variability of renewable electricity sources: an updated review. Energy Sust Dev 2010;4:302—14.

[61] Leach A, Doucet J, Nickel T. Renewable fuels: policy effectiveness and project risk. Energy Policy 2011;39(7):4007—15.

[62] Agrawal A. Risk mitigation strategies for renewable energy project financing. Strat Plann Energy Environ 2012;32(2):9—20.

[63] Cucchiella F, D'Adarno I. Feasibility study of developing photovoltaic power projects in Italy: an integrated approach. Renew Sustain Energy Rev 2012;16(3):1562—76. https://doi.org/10.1016/j.rser.2011.11.020.

[64] Rangel LF. Competition policy and regulation in hydro dominated electricity markets. Energy Policy 2008;36(4):1292—302. https://doi.org/10.1016/j.enpol.2007.12.005.

[65] Nikolic DM, Jednak S, Benkovic S, Poznanic V. Project finance risk evaluation of the Electric power industry of Serbia. Energy Policy 2011;39(10):6168—77. https://doi.org/10.1016/j.enpol.2011.07.01.

[66] Berchum EV. Pumped hydro storage pressure cavern large-scale energy storage in underground salt caverns (Master's thesis). Retrieved from. TU Delft Repositories; 2014.

[67] Hendriks M. Local hydroelectric energy storage. A feasibility study about a small scale energy storage system combining. hydropower, gravity power, spring power and air pressure (Master's thesis). Retrieved from. TU Delft Repositories; 2016.

[68] Taw. Leidraad kunstwerken. 2003. http://www.helpdeskwater.nl/publish/pages/5175/ad012-leidraadkunstwerken.pdf.

[69] RIVM. Handleiding risicoberekeningen bevi. 2009. Retrieved from, http://www.rivm.nl/dsresource?objectid=rivmp:22449&type=org&disposition=inline&ns_nc.

[70] Gary L. Managing project risk. Harvard Business School Publishing Corporation; 2005. 3.

[71] Weron R. Energy price risk management. Physica A 2000; 285:127–34.

[72] Nicolić Makajic D, et al. Project finance risk evaluation of the Electric power industry of Serbia. Energy Policy 2011; 39:6168–77.

[73] Locatelli G, Invernizzi DC, Mancini M. Investment and risk appraisal in energy storage systems: a real options approach. Energy 2016;104:114–31.

[74] Sovacool BK, Gilbert A, Nugent D. Risk. Innovation, electricity infrastructure and construction cost overruns: testing six hypotheses. Energy 2014;74(September): 906–17.

[75] Ericson II CA. Hazard analysis technique for system safety. John Wiley & Sons; 2005.

[76] Ayyub BM. Risk analysis in engineering and economics. Chapman and Hall/CRC; 2003.

CHAPTER 4

Gravity Energy Storage Applications

INTRODUCTION

Energy production is expected to change considerably in the upcoming decades. The growing scarcity of currently used energy resources, environmental concerns, and energy security are the main drivers for the transition toward renewable energy systems. However, the increasing use of these intermittent energy resources is pressing need for energy storage. The energy transition from conventional power plants to clean energy generation alternatives necessitates more flexibility at different electricity value chain including generation, transmission, distribution, and consumption. This operation flexibility can be provided by energy storage to obtain a stable and robust power system. Energy storage operated in small- and large-scale applications enable today's power system to operate efficiently with more power reliability and lower price. Various beneficial grid services and cost-saving functionalities are offered by energy storage technologies. These systems are being deployed by utilities and customers for several diverse purposes. Typically, conventional energy sources do not operate at their peak performance and are turned on and off following energy demand fluctuations. The use of traditional energy sources to meet our energy needs is costly and not environmentally friendly. Additionally, these generation power plants are not able to respond to real-time increase in energy demand because of their slow ramp-up time, which could lead to poor power quality.

The widespread penetration of renewable energy systems in the electric grid has created an equal need for energy storage technologies. The availability of solar and wind energy depends on the weather and cannot be commanded. Energy generation from these sources does not match demand. Electric utilities have to deal with the increasing use of distributed energy sources such as photovoltaic (PV) systems installed at residences that may not be producing power during peak demand periods. For instance, customer cannot run a number of equipment including an air conditioner in the absence of PV production during the unavailability of sunlight. Excess energy is stored in energy storage systems to be delivered later for consumption to ensure a reliable and continuous power provision throughout the day. The storing of energy for use at a different period is considered one of the many services and applications of energy storage systems. These systems are also responsible for regulating frequency with an aim to improve power quality. Furthermore, energy storage allows owners to make profit by selling energy when it is at its highest costs, as well as ensuring the delivery of uninterruptible power to critical services.

ENERGY STORAGE SERVICES

Energy storage has been considered of great interest to electric utilities for a long time because of the potential functionalities they offer to support the electric grid. Traditionally, load leveling was considered one of the most important services provided by energy storage as it enabled the reduced use of expensive peak energy generation systems. With the high integration of renewable energy systems, this service has been extended to include other functionalities to support the intermittent nature of renewable energy sources. Electric utilities have recently started considering energy storage as an alternative to power grid system upgrade as it contributes to the optimization of its infrastructure and hence defers the development and installation of new electric power lines. In addition to that, energy storage offers other technical functionalities known as ancillary services.

The continued interest in the development and deployment of energy storage systems is driven by the growing importance of the potential services provided by these systems. A variety of benefits are provided to the electric power grid with the use of energy storage systems. These functionalities can be classified into the following:
- Energy supply
- Power grid operations
- Grid infrastructure
- End user
- Renewable energy integration

Energy storage has become an important component of the traditional electricity value chain, which consists of energy source, generation system, transmission and distribution (T&D) system, as well as end-user side. Because of the rich spectrum of services it provides for the aforementioned modules of the electricity value chain, energy storage has created a more responsive energy market. Some important applications have been summarized in Refs. [1–4].

Energy Supply

The benefits associated with electric supply provided from energy storage include energy time-shift and energy supply capacity. The first one is defined as buy low-sell high.

Energy time-shift

This service involves the charging of the storage system when energy prices are low; the stored energy is later sold at higher values during peak energy demand. An illustration of energy time-shift provided by energy storage is shown in Fig. 4.1.

The objective behind the use of energy storage for time-shift application is the utilization of low-priced electricity during periods of high energy prices. During peak demand, the cost of energy production is high due to the use of peaking power plants. Therefore, the stored energy comes from baseload generation systems such as combined cycle plants whose production of energy has to remain constant, from wind plants whose generation outputs occur during periods of low energy demands, or from energy generation systems whose incremental cost of energy production is low such as hydroelectric and geothermal power plants [4].

Electric power utilities may use electric time-shift service to decrease energy-related cost driven by reduced cost and need for generation fuel. This service may be used also by commercial owners of energy storage to make profit from buying low-cost wholesale energy and selling it at higher prices.

Electric supply capacity

Providing energy supply capacity is another service offered by energy storage. Electric supply capacity is reduced by energy storage discharged power. The main objective of this storage functionality is to reduce the need for power generation equipment. The deferred energy supply capacity resource includes expensive and less efficient combustion turbines, combined cycle generators, natural gas, and coal baseload generation.

Electric power utilities may use energy storage to provide electric supply capacity to decrease capacity-related costs. This service may be used by owners of energy storage plants to make profit in a capacity market.

Grid Operation

The use of energy storage enhances grid operation by providing what is known as ancillary services. These services are defined by the Federal Energy Regulatory Commission (FERC 1995) as functionalities necessary to maintain and support the operation of the transmission system in a reliable manner.

Energy storage systems are well positioned to provide various ancillary services needed for a reliable and stable electricity grid. The provision of ancillary services by energy storage reduces the use of other generation systems and fuel and decreases air emissions.

Load following

During period referred as peak hours or shoulder hours, energy storage provides load following. Load following up is offered by discharging more energy from the storage, whereas load following down is performed by increasing the charging of the system. An increase of energy demand requires energy storage to provide load following up by increasing the discharged energy. Conversely, a decrease of energy demand leads to a reduction of generation output to provide load following down. Load following service offers a number

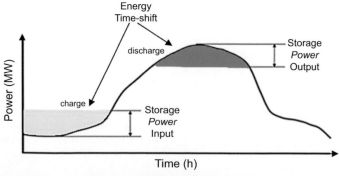

FIG. 4.1 Electric energy time-shift.

of benefits such as reducing the need for other energy generation systems, energy production variability and fuel use, as well as negative environmental impacts (air emissions).

Frequency regulation

Frequency regulation, also known as area regulation, is an ancillary service whose aim is to match moment-to-moment energy demand and supply. The main objective of this service is to maintain the stability of alternating current frequency within a certain area. An illustration of this service is shown in Fig. 4.2. During excess momentarily energy supply, frequency regulation down is required to balance the demand with the supply. Contrariwise, frequency regulation up is necessary when energy supply is momentary less than energy demand.

The significant penetration of renewable energy systems in the electric grid such as wind and solar energy technologies will cause energy generation output to vary along with energy demand. Frequency regulation is necessary to offset this variation in supply and demand as illustrated in Fig. 4.3.

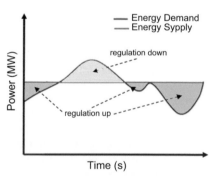

FIG. 4.2 Frequency regulation.

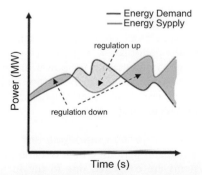

FIG. 4.3 Frequency regulation up and down.

Frequency regulation service is offered by energy storage as similar to load following. Area regulation up is provided by increasing the discharging of the storage system and/or decreasing its charging. Conversely, area regulation down is performed by decreasing the discharging of the storage and/or increasing its charging. Energy storage technologies with high energy efficiency are mostly preferred to provide frequency regulation service.

Frequency response

Frequency response is an ancillary service, which can be provided by energy storage systems with a very fast ramp rate. The role of storage is to control the frequency and respond to anomalies over the time span of milliseconds. The purpose of this service is to maintain frequency close to the targeted one. The difference between area regulation and frequency response is that the first service responds indirectly to frequency with the use of control signals which reflect the variation between energy demand and supply, whereas the second one controls AC frequency directly. In addition, the response time of the aforementioned services is different. That is, the variation of output from resources used to provide frequency response should be faster than the output variation of other resources performing area regulation service. Only few systems are characterized by a fast ramp rate, among them energy storage. Fast storage systems are perfectly suited for frequency response application. The use of these devices in the power grid offsets the need for quick response generation resources.

Ramping

Grid stability is affected by ramping if this later becomes significant. Power system operators have to deal with this challenge to ensure the stability of the grid. Ramping refers to high changes of energy output ranging from few seconds to minutes. A good example could be variation in wind power generation due to quick changes in wind speed, which results in ramps up or down outputs. The impact of ramping increases as more renewable energy generation systems are integrated into the grid.

Resources involved in ramping services should be able to counterbalance output ramping similar to area regulation and load following. These resources should hence be capable of providing energy output variability by increasing or decreasing output to match changes in energy generation.

Most conventional generation resources are not very well suited for this service as they should be characterized by a rapid varied output. Power system operators

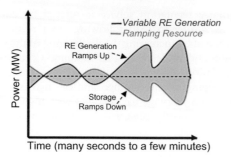

FIG. 4.4 Ramping up and down.

may operate supplementary combustion turbines to offset additional needed generation capacity.

Energy storage is an interesting solution that could be used for ramping services. Storage technologies provide both ramping up and down options. By reducing storage charging or/and increasing the discharging of this, energy storage offers ramps up. Inversely, by increasing storage charging or/and decreasing its discharging, the system offers ramping down service. Fig. 4.4 illustrates storage ramping service.

The utilization of energy storage systems in ramping functionalities results in a reduction in a number of aspects, such as energy production variability, generation start-ups, need for generation capacity, and fuel use.

Reserve capacity

Reserve capacity is a backup energy generation capacity that is used by the electric grid in the occurrence of unexpected fault such as the unavailability of a power plant. Energy storage systems have the ability to provide this service and are used to offset or reduce costs incurred for generation of reserve capacity. This service has three categories which include the following:

- **Spinning reserve**: Also referred as synchronized reserve, this type of reserve capacity is the first one used during the occurrence of a shortfall. It is an unloaded online generation capacity used for compensation of transmission or generation outage. It has a response time of 10 min.
- **Supplemental reserve**: This type of reserve capacity is used after spinning reserve. It may be an offline generation capacity, which can respond within 10 min.
- **Back supply**: It is considered as a backup for both supplemental and spinning reserves.

Voltage support

Maintaining the stability and the required voltage level of the electric grid is the most challenging technical work of grid operators. These aspects are achieved through proper management of reactance at the grid level. Electric grid operators make use of voltage support, which is an ancillary service to manage grid reactance. Historically, this service has been provided by generators, which produce reactive power. Recent alternative systems which offer voltage support include new technologies such as power electronics, energy storage, as well as control and communication systems.

Distributed energy storage systems located close to end users are well suited for this grid application and have gained great interest. The main reason behind the use of distributed storage for such service is because reactive power is not transmitted effectively over long distances. Therefore, voltage support is well provided by distributed storage located in regions where most reactance happens.

Black start

Back start system refers to units able to energize the electric grid after an outage. These systems are capable of starting up on their own without the provision of power from the grid. They operate on standby mode until they are called upon. Most energy storage systems are able to provide this service and are classified as black start resources due to their ability to operate without a need for any special equipment.

Grid Infrastructure

Energy storage is expected to play an important role not only in the power grid but also as an important asset of the utility T&D system because of its modularity, flexibility, and operational characteristics.

The benefits of grid infrastructure come from the utilization of energy storage within the T&D utility system. Energy storage increases the performance of T&D facilities, improves their carrying capacity, and develops their reliability. Furthermore, energy storage can be used to avoid T&D congestion, extend the life span of T&D equipment, and defer the upgrade and the use of additional T&D capacity and equipment.

Transmission support

Energy storage systems are used to support transmission by improving the performance of the T&D system. This is done through compensating for power disturbance and anomalies, such as voltage instability and sags. Transmission support benefits are highly dependent on the site and its location.

Transmission congestion management

During peak demand periods, a high number of transmission systems are congested due to the increasing use of distributed energy resources and renewable energy generation. The addition of transmission capacity does not keep up with the growing deployment of renewable generation, which results in charges associated with transmission capacity congestion. To avoid these charges, utilities or end users should make use of energy storage system. This can be sited near the congested part of the transmission system to reduce transmission capacity congestion. In the absence of transmission congestion, energy storage is used to store energy. During peak demand, load is served by the stored energy, which is discharged from the storage system, hence, reducing the need for energy, which must be provided by the transmission system.

Benefits received from providing transmission congestion management by the use of energy storage are based on avoided congestion charges. It is important to note that energy storage providing management of transmission congestion can also results in increased energy transmission annually (kWh per kVA of transmission capacity) if a substantial amount of energy is sent to storage during off-peak demand periods.

Transmission and distribution upgrade deferral

Upgrade investments in T&D system can be delayed or avoided by the use of energy storage. This energy storage service is known as T&D upgrade deferral. Fig. 4.5 shows an illustration of this service. The rated carrying capacity of the T&D system and the energy demand are shown in Fig. 4.5. It can be seen that at a specific time of the day, peak demand surpasses the load-carrying capacity of the T&D system. A typical solution to avoid this problem is increasing the T&D load-carrying capacity few years before the occurrence of this expected overload. It should be noted that extra capacity cannot be practically added to the T&D system.

Rather, the existing equipment have to be replaced by equipment with higher rating capacity. Another alternative could be the addition of many equipment to increase the capacity of the existing ones. Usually, a 33%–50% increase of capacity is used [4].

The use of energy storage as an alternative to T&D upgrade has gained attention in recent years. Energy storage systems are placed downstream the T&D overloaded equipment to lessen peak demand, which has to be provided by the aforementioned equipment. The benefits received from performing this T&D service are interesting and can be in the range of hundreds of dollars per kW per year.

More details about T&D upgrade deferral benefits are found in a study conducted by Sandia National Laboratories [5].

Transmission and distribution equipment life extension

Very much like T&D upgrade deferrals service, T&D equipment life can be extended by the use of energy storage in the grid. By reducing loading, energy storage can reduce the existing equipment wear and extreme heating, thus, extending their expected useful life span. For example, the use of storage can extend the life of underground distribution cables by decreasing their peak loading and hence
- reducing the insulation degradation of the cable and
- reducing the occurrence of ground faults, which may have a negative effect on the cable lifetime.

This is an attractive storage service especially when equipment are located in populated and developed regions characterized by high replacement costs. For instance, replacement would cause disruption in some urban areas and would necessitate an expensive construction permit.

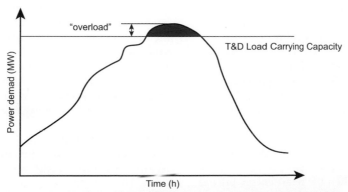

FIG. 4.5 Transmission and distribution load-carrying capacity growth.

Benefits received from performing equipment life extension and T&D upgrade deferral are compelling because energy storage systems enable utilities to avoid or defer an expensive investment in new equipment.

This high investment could be delayed especially if the occurrence of the highest loads on a T&D system node is only few hours per year. Additionally, the use of energy storage is more attractive when it is located in areas with uncertain load increase. For instance, it is necessary to upgrade the existing T&D equipment in regions where large residential buildings are expected to be built to accommodate the increased demand. In case there is uncertainty about the project completion time or load growth, it is more interesting for utilities to use energy storage to delay equipment investment until this information is certain.

On-site power

A high number of electrochemical batteries are owned by electric utilities for the provision of on-site power back up at substations. During the unavailability of grid power, energy storage systems are operated to deliver power to control and communication equipment. Utility substations in the United States have more than 100,000 battery storage systems used for this purpose [4]. Most of these are vented lead-acid battery type.

End User

Energy storage owned by end users could be used for a number of applications. Customers can use energy storage to control their energy bills and make profit through power purchase agreement for ancillary services, energy, and capacity in the spot market. Likewise, power merchants can aggregate on-site storage systems for the provision of other applications. A number of benefits can be received from the use of end-user sited energy storage system such as management of time of use (TOU) energy cost and demand charge, as well as electric service reliability and power quality. The two first services enable the end user to manage his bill by reducing TOU energy cost and demand charges. Whereas the two other services are complementary as they allow customer to ensure reliability and power quality of the electric service and hence avoid costs.

Cost management

Cost associated with TOU energy can be reduced by customers with the use of energy storage. This can be done by charging the storage when the retail energy price is low during off-peak energy demand. The stored energy is used during periods of high energy demand when its price is high. This storage application is very similar to the application of energy time-shift, also known as arbitrage.

The qualification of energy end uses depends on the type of retail tariff involving energy prices, which represent time-specific rates. Tariffs for energy TOU include rates that are particular to energy time-of-day, day of week, and season (usually summer and winter).

Demand charge management

An interesting energy storage application for end users is demand charge management. The objective of this application is the reduction of energy demand with an aim to offset or avoid peak energy demand charges. Utility tariffs for commercial end users include distinct charges for power and energy; that is why the opportunity of managing demand charge exists. Demand charges (power-related) are evaluated based on the maximum power used by the customer.

Similar to TOU energy, energy charges are also particular to time of day, season, and day of week. Usually, there exist a number of demand periods, classified into three to five ones such as super on-peak, on-peak, mid-peak, off-peak, and super off-peak.

The highest demand charges occur during periods of superpeak and peak demand. For a typical summer weekday, the aforementioned periods are between 10 a.m. and 7 p.m. Conversely, lower demand charges occur during off-peak or mid-peak periods. Typically, these appear in summer weekdays in the morning and evening, weekend days in summer, and winter months during weekdays.

The unit used to express demand charges is $/kW-month. That is, demand charges are evaluated monthly depending on the month maximum demand. To avoid demand charges within a given month, demand should be minimized during all peak-demand periods.

Energy is stored during periods of low demand charges to reduce energy purchased from the grid when demand charges are high. Similarly, at low demand charge, energy price for charging the storage is also cheap. The energy stored is then used during periods when demand charges apply. The storage discharged energy would serve the end-user load and avoid the need to buy energy from the electric grid.

Electric service reliability

Energy storage systems are commonly used to avoid electricity interruption and ensure electric service reliability. The role of energy storage in this application is similar to uninterruptible power supply (UPS).

In the occurrence of power outage, energy storage is discharged to avoid outage.

Electric service power quality

This application is similar to the abovementioned storage service, which is about electric reliability. Instead of ensuring reliability, energy storage systems are used to protect equipment from the impact of grid poor power quality. Power quality issues have become more prevalent and occur during short periods. The main causes of poor power quality include variation of voltage (dips, sags, surges, or spikes) and electrical noise that is the occurrence of oscillations and high frequency transients.

The benefits received from the use of energy storage in electric service power quality are based on avoided charges associated with damage of equipment, substandard equipment operation, and equipment downtime.

Renewable Energy Integration

Energy storage systems are expected to play an important role in the integration of renewable energy technologies into the future electric grid. There is a significant need for the use of energy storage to overcome the variable output of renewable energy systems and to ensure their reliable and effective integration. The most important issues that will be addressed by energy storage include power output variability, mismatch between energy generation and demand, weather forecasts uncertainty, and undesirable electrical impact on the grid.

The intermittent energy production output of renewable energy systems depends on the duration period of this variability, which could be classified as a short or long duration. The first category is mainly caused by the inconsistency of wind speed or the prompt variability of solar energy output because of clouds. Short variability duration typically lasts for few seconds to several minutes. This rapid change of generation output needs to be offset by energy storage. Various types of energy storage systems are perfectly suited to provide ramping services to reduce variation of energy generation output. The second category occurs for a longer period of time, which could be from day to day, year to year, or season to season. Energy storage is also used to address the long duration variability issue. The storage system discharges energy when renewable energy systems are not generating full power.

The lack of accuracy and uncertainty of weather forecasts is another challenge associated with renewable energy generation. The occurrence of unexpected shortfalls of renewable energy output due to imprecise weather forecasts should be addressed by other resources to correct this difference. Energy storage is well suited to overcome this challenge by providing back up energy when needed.

Renewable energy generation can have some undesirable impact on power quality of the electric grid. For instance, the high integration of solar PV can lead to voltage variability, which will affect power quality of the overall electric grid. Energy storage could be used to address several power quality issues resulting from renewable energy integration. In this manner, the storage system can absorb excess or abnormal power in case the variability of voltage on a local distribution system is high. To improve grid voltage, energy storage provides both real and reactive power. Furthermore, in the occurrence of undesirable backflow, energy storage can absorb excess power to reduce this later.

Renewable energy systems and energy storage are considered somewhat complementary as they have a number of synergies. The variable renewable energy output can be offset by energy storage. The use of this in the electric results in a number of benefits such as the following:

- Reducing the use of conventional energy generation systems.
- Avoiding power quality anomalies.
- Enabling the integration of variable RE systems.
- Reducing the use of a number of power-conditioning equipment.
- Increasing the value of energy produced by RE technologies.

Financial Benefits of Energy Storage Systems

The different financial benefits that could be obtained from the use of energy storage are summarized as follows:

Increase of revenues by the use of bulk energy arbitrage

This service is about charging the storage system during periods of low energy demand by purchasing cheap excess electricity. The stored low-priced energy is then sold during periods of high electricity prices.

Increase of revenues by centralization of generation capacity

The use of energy storage in regions characterized by a tight energy generation capacity is financially beneficial and can be used to offset the cost of installing new-generation systems or to avoid the renting of generation capacity in the wholesale marketplace.

Increase of revenues by the provision of ancillary services

Energy storage systems offer a number of ancillary or support services to the electric grid to ensure its proper operation such as load following, regulation, and reserve capacity.

Reduction of transmission congestion costs

The performance of the T&D system is improved by the use of energy storage. This ensures voltage stability and provides utilities with the ability to increase the transfer of energy. Therefore, charges resulting from transmission congestion are avoided by the use of these technologies in the electric grid.

Reduction of energy demand-related costs

By reducing the end-consumer use of energy during peak periods or when energy prices are high, demand charges are reduced. This could be achieved by the use of energy storage.

Prevention of energy reliability-related charges

Power outage associated costs can be avoided by the utilization of energy storage. This financial benefit concerns mainly industrial and commercial end users, which could be significantly affected by power outages.

Reduced financial losses related to power quality

Power quality issues results in financial losses that could be avoided by the use of energy storage. Such power quality anomalies have negative effects on loads and can cause equipment damage.

Increase of revenues by the use of renewable energy systems

Energy produced from renewable energy sources could be time shifted by energy storage. During periods of low energy demand and electricity prices, energy can be stored. The stored energy could be used when renewable energy generation is low and when electricity price and demand are high.

The aforementioned financial benefits point out that energy storage has various sources of revenues due to the different functionalities it performs, which should be considered when investigating the profitability of energy storage technologies in combination with renewable energy systems.

Mapping of Energy Storage with Grid Applications

The most installed energy storage worldwide regarding system power rating is pumped hydroenergy storage (PHES) because of its simplicity and resemblance to the well-established hydroelectric power systems. Pumped hydrostorage (PHS) is properly suited for applications requiring long discharge time and high-power rating such as balancing of renewable energy intermittency. Compressed air energy storage (CAES) can also be used in a number of similar applications as PHS. Unlike PHS, CAES is characterized by less power capacity and shorter discharge time than PHS, which enables the system to perform other additional services such as power quality and reliability. The main issue facing the aforementioned technologies is the requirement for specific geographic construction sites. This has led researches to investigate the development of alternative energy systems based on the well-developed principle of PHS such as gravity energy storage (GES). This technology is expected to perform services analogous to the ones performed by PHS and CAES because of the similarity of its system characteristics with its counterparts. Flywheel energy storage systems are more appropriate for spinning reserve services, load following, and power capacity as they are characterized by lower capacity and faster discharge time. Superconducting magnetic energy storage is also suitable for power quality. Batteries can be used for several storage applications as shown in Table 4.1. Unlike other storage systems, batteries can provide off-grid applications and can also be used in transportation. The current transition toward electric vehicles would results in a significant use and production of batteries such as lithium-ion. This will positively impact the battery industry leading to an improved battery system with a reduced price.

The applications of hydrogen fuel cells are limited compared with batteries. The main discrepancy between them is the conversion efficiency. Hydrogen systems are characterized by a low conversion efficiency from electricity to hydrogen and then backward. This type of energy storage cannot be implemented to provide a number of grid services, such as spinning reserve and frequency regulation.

The future expected trend would be probably the use of PHS, CAES, and GES for bulk storage applications. Inversely, flywheels are not considered a site-depend storage option and would be used to provide other services not served by the aforementioned mechanical energy storage systems. Lithium-ion batteries are predicted to play a significant role and contribution in most energy storage applications. Conversely, NaS batteries are most likely to take part of the future electric grid due to its cheap cost and higher lifetime compared with some other batteries.

TABLE 4.1
Applications of Energy Storage Systems [6].

	Mechanical Energy Storage (ES)				Electrochemical ES				Electrical ES	Hydrogen ES	Thermal ES
	PHES	CAES	FES	GES	Lead-Acid	NaS	Li-Ion	Flow Battery	SEM	Fuel Cell	TES
Energy arbitrage	✓	✓		✓				✓		✓	✓
Load following	✓	✓	✓		✓	✓	✓	✓			✓
Peak shaving	✓	✓		✓	✓	✓	✓	✓			✓
Voltage support			✓		✓	✓	✓	✓			
Spinning reserve	✓	✓	✓		✓	✓	✓	✓			✓
Frequency regulation	✓	✓		✓	✓	✓	✓	✓			
Black start	✓	✓		✓	✓	✓	✓	✓			
Power quality		✓	✓		✓	✓	✓	✓	✓		
Power reliability		✓	✓		✓	✓	✓	✓			
Renewable energy application	✓	✓		✓	✓	✓	✓	✓		✓	✓
Off-grid application					✓	✓	✓			✓	
Transportation application					✓	✓	✓			✓	

Energy Economics
Present value

There exist a number of methods to determine the economic viability of an energy storage project. The system capital cost, replacement cost, along with operation and maintenance costs have to be combined and compared with the benefits received from doing the project. This analysis entails taking into consideration the time value of money, which refers to the concept that 1 dollar on hand today is much greater than 1 dollar received in the future. A present worth analysis has to be conducted to account for these differences. That is, all future costs should be converted to present value equivalence. The relationship between the present worth (P) of a future amount of money (F) is expressed as:

$$P = \frac{F}{(1+i)^n} \qquad (4.1)$$

Typically, benefits received from a project investment may be annually. In this case, a conversion factor called present value function (PVF) is used to determine the present value of a stream of cash flows delivered annually, with an interest rate (discount rate) (i) and for a number of years into the future (n).

$$P = A.PVF(i, n) \qquad (4.2)$$

PVF is the sum of the present values for a stream of n annual, which start the first year from the present. PVF is given as:

$$PVF(i, n) = \frac{1}{1+i} + \frac{1}{(1+i)^2} + \dots + \frac{1}{(1+i)^n} \qquad (4.3)$$

Eq. (4.3) yields to

$$PVF(i, n) = \text{Present value function} = \frac{(1+i)^n - 1}{i(1+i)^n} \qquad (4.4)$$

Payback period

The payback period is the ratio of the project cost to the annual received savings (Eq. 4.5).

$$\text{Simple Payback} = \frac{\text{Extra First Cost } \Delta P(\$)}{\text{Annual Saving } S(\$/yr)} \qquad (4.5)$$

Rate of return

The rate of return is the inverse of the payback period. That is, it is the ratio of the annual received savings to the investment cost (Eq. 4.6).

$$\text{Interest rate of return} = \frac{\text{Annual Saving } S(\$/yr)}{\text{Extra first cost } \Delta P(\$)} \qquad (4.6)$$

PROFITABILITY MODELING IN SMALL- AND LARGE-SCALE APPLICATION

To evaluate the profitability of an energy storage technology, it is vital to take into account all the different scenarios in which the storage system could be used for. Energy storage systems could be integrated to the grid or connected to a stand-alone system for an off-grid application. Both small- and large-scale applications of GES would be investigated in this study. The energy demand would be represented by a residential house for a small-scale application, whereas the demand side for a large-scale application would make use of energy demand in a particular city. The energy generation resource would be represented by a PV system or wind turbines.

Problem Formulations

A model is developed in this section to investigate the economic profitability of GES. The purpose of this mathematical problem is to determine the benefits delivered from optimally operating the storage system. The objective function of the model is to maximize the revenues of the system owner by performing arbitrage, and/or grid service while taking into account the operational constraint, the demand side, and the variable energy market prices. A linear programming model is developed and solved using General Algebraic Modeling System (GAMS). The model input variables are the hourly energy price, PV energy generation, and the different characteristics of GES. These include system lifetime, rated power, energy efficiency, and energy storage capacity. The model outputs include the hourly project revenues and the hourly exchanged energy between the system components.

The net present value (NPV) is calculated to determine the profitability of the system by comparing the present value of the benefits delivered with all incurred costs. That is, all cash flow steams have to be converted to present worth to obtain an adequate comparison. The expression used for the calculation of the NPV is given by (Eq. 4.7):

$$NPV = \sum \frac{(\text{Benefits} - \text{cost})^n}{(1+i)^n} \qquad (4.7)$$

where i is the discount rate and n is the analytic horizon (in years).

The benefits of installing energy storage in an electrical system are identified by calculating the revenues received from the use of this energy storage minus the revenues received without its use (Eq. 4.8). This methodology is used to capture the benefits received only from the use of storage system in a grid application.

The objective of this model is to maximize the benefits from optimally scheduling the selling and the purchasing of energy to optimally operate the system. An estimation of the aforementioned revenues along with storage costs is provided in the following subsections.

$$\text{Benefits} = \text{Rev}_S - \text{Rev}_{WS} \quad (4.8)$$

Revenues Model
Revenues obtained from the use of energy storage
The model is formulated by taking into account all the energy exchange variables between the electric grid, the storage, the load, and the generation resource, which is represented by a PV system in this case.

$$\text{Rev}_S = \sum (E_{Sold}(t) - E_{Purchased}(t))E_P(t) \quad (4.9)$$

The energy sold to the grid $E_{Sold}(t)$ is the sum of the energy injected to the grid from the storage $E_s^g(t)$ with the energy sent to the grid from the PV system $E_{PV}^g(t)$. Energy sold is expressed as Eq. (4.10):

$$E_{Sold}(t) = E_s^g(t) + E_{PV}^g(t) \quad (4.10)$$

The energy purchased from the grid $E_{Purchased}(t)$ is the sum of the energy bought from the grid to charge the storage $E_g^s(t)$ with the energy transferred from the grid to the load $E_g^L(t)$. Energy sold is expressed as Eq. (4.11):

$$E_{Purchased}(t) = E_g^s(t) + E_g^L(t) \quad (4.11)$$

The hourly energy demand $E_{Demand}(t)$ is met by transferring energy from the storage $E_s^L(t)$, the grid $E_g^L(t)$ and the PV system. Energy demand is given by (Eq. 4.12):

$$E_{Demand}(t) = E_s^L(t) + E_g^L(t) + E_{PV}^L(t) \quad (4.12)$$

The energy generated by the PV system $E_{PV}(t)$ is supplied to the load $E_{PV}^L(t)$, or/and used to charge the storage system $E_{PV}^s(t)$, and/or injected (sold) to the electric grid $E_{PV}^g(t)$, as formulated in Eq. (4.13):

$$E_{PV}(t) = E_{PV}^L(t) + E_{PV}^s(t) + E_{PV}^g(t) \quad (4.13)$$

The storage system is charged either from the solar PV or the grid. Conversely, the energy discharged from the energy storage system is supplied to the load or

sent to the grid. The charging and discharging states of the system are presented in Eqs. (4.14)–(4.15), respectively.

$$E_{Stored}(t) = E_{PV}^s(t) + E_g^s(t) \quad (4.14)$$

$$E_{Discharged}(t) = E_s^L(t) + E_s^g(t) \quad (4.15)$$

Revenues obtained without the utilization of energy storage
Because the system does not make use of energy storage, energy is exchanged with the grid, the load, and the generation source. The equations used by this model are similar to the above presented formulas, except that they do not incorporate energy exchange variables associated with energy storage. The following equations (Eqs. 4.16–4.20) are used in the calculation of revenues of the system without the use of energy storage.

$$\text{Rev}_{WS} = \sum (E'_{Sold}(t) - E'_{Purchased}(t))E_P(t) \quad (4.16)$$

$$E'_{Sold}(t) = E_{PV}^g(t) \quad (4.17)$$

$$E'_{Purchased}(t) = E_g^L(t) \quad (4.18)$$

$$E'_{Demand}(t) = E_g^L(t) + E_{PV}^L(t) \quad (4.19)$$

$$E'_{PV}(t) = E_{PV}^L(t) + E_{PV}^g(t) \quad (4.20)$$

As mentioned before, the benefits received are discounted over the years of the project. The future annuity cash flows are converted to an equivalent present value with the use of the present value factor (Eq. 4.1). The model considers a number of constraints related to energy storage as it would be discussed in the following subsection. In addition, energy transaction parameters must be positive.

Energy storage constraints
The storage state changes with time because of the discharging and the charging of the system. The energy storage level should be taken into account when formulating the operation model. The storage hourly state (Eq. 4.21) is formulated by taking into account the system self-discharge (δ) and energy efficiency (μ), as well as the energy charged and discharged from the system.

$$S_L(t) = (1 - \delta)S_L(t-1) - E_{Discharged}(t) + (E_{Stored}(t)\mu) \quad (4.21)$$

The storage system level at a specific time interval takes into consideration the remaining energy at (t-1), the energy that flows into the storage at that time, and the energy discharged from the system at time t.

The state of the storage is restricted by the maximum storage capacity (E_{max}) and zero. Furthermore, the system energy level has to be greater than or equal to the system discharged energy as expressed in (Eq. 4.22).

$$0 \leq E_{Discharged}(t) \leq S_L(t) \leq E_{max}(t) \qquad (4.22)$$

Another storage constraint is related to the system power and capacity limits. The energy that flows in (stored) and out (discharged) from the storage system has to be less than or equal to the following limits (Eq. 4.23):

$$E_{Discharged}(t) \leq E_L \text{ and } E_{Stored}(t) \leq E_L \qquad (4.23)$$

Energy Storage Cost

Energy storage total cost is a combination of several costs including energy and power capacity, operation and maintenance, replacement, as well as balance of plant cost. These costs are dependent on the storage technology. The energy storage capacity and power costs are the main storage cost. The balance of plant cost includes permit, site fees, taxes, and construction-associated costs. The operation and maintenance cost consists of fixed annual cost ($/kW) and variable cost ($/kWh). For the proper operation of the system, an annual maintenance is required; this is considered a fixed annual cost. The variable cost depends on the storage delivered energy; this cost is very low compared with the other storage cost and can be neglected for some storage systems. The energy storage cost comprises the system capital cost, the operation and maintenance cost, as well as the system replacement cost. The present value of these costs is formulated as:

$$P(Cos\ ts) = C_S + O\&M + R_p \qquad (4.24)$$

The system capital costs include the storage power and capacity costs, as well as the balance of plant (BOP) costs (Eq. 4.25):

$$C_S = P_c + C_c + BOP \qquad (4.25)$$

where

$$\begin{cases} P_c = U_P \cdot P_O \\ C_c = \dfrac{U_c E}{\mu} \\ BOP_c = U_{BOP} \cdot P \end{cases} \qquad (4.26)$$

The storage system operation and maintenance costs are calculated using (Eq. 4.27):

$$O\&M_c = U_{O\&M} \cdot P \cdot \mu \qquad (4.27)$$

The replacement cost of the storage system is expressed as (Eq. 4.28):

$$R_P = R \sum_{d=o}^{d} (1+i)^{-dn'} \qquad (4.28)$$

Development of Scenarios

The proposed model is solved while considering all the different applications of energy storage. A number of scenarios for residential application are determined and classified into two categories. The first group deals with GES connected to the electric grid, whereas the second one represents off-grid storage system coupled to a generation source. The revenues of the first category are received from performing arbitrage service, whereas the second-type benefits from T&D deferral service. Fig. 4.6 shows an illustration of the system.

The first category that deals with GES connected to the grid includes three scenarios. The first and the second scenarios include an energy storage, a load, and an electric grid. The third scenario of the first category makes use of a generation source in addition to the components used in the two aforementioned scenarios.

The second category has only one scenario in which gravity storage coupled with a PV is used as a stand-alone system.

In the first scenario of the first category, GES is connected to the grid and not the load. Hence, energy demand is met by the grid. The energy purchased from the grid is used either to charge the storage or to supply the demand side. On the other hand, energy is sold to the grid from discharging the storage. Energy storage is used to provide arbitrage service and makes profit by optimally purchasing and selling energy. In this case, the energy exchange variables $(E_s^L(t)E_{PV}^L(t),$ $E_{PV}^g(t)E_{PV}^s(t))$ are equal to zero.

Similar to the first scenario, the storage in the second scenario is connected to the utility grid. In this situation, the demand side is supplied energy from the storage and the grid. The model mathematical variables $(E_{PV}^L(t)E_{PV}^g(t)E_{PV}^s(t))$ are equal to zero because of the absence of a generation source (PV).

A hybrid system composed of a PV system with GES is used in the third scenario. Energy is supplied to the demand side from the PV system, the storage, and the electric grid. The main purpose of energy storage in this situation is to optimally balance the demand and the supply of energy with an aim to increase profit.

The second category deals with off-grid storage. This last scenario consists of a load, a PV system, and gravity storage. The hybrid system is responsible for supplying

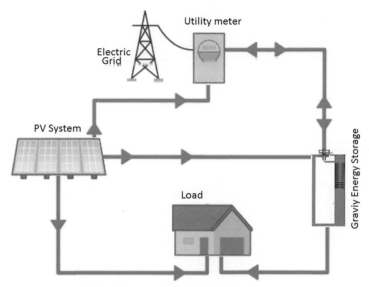

FIG. 4.6 Schematic of the studied system.

energy to the load. Profit is made from performing T&D. This storage service enables utilities to delay or offset charges associated with T&D upgrade. Based on literature review, a rough estimation of this service has been used. The benefits of this service are quantified as between $150,000 and $1,000,000/MW-year [7]. This analysis makes use of the average to avoid over or underestimating its value.

Case studies for the four discussed scenarios of GES for both small- and large-scale applications are analyzed to evaluate the effectiveness of the presented models.

Case Study for Residential Small-Scale Application

The input parameters of the model include hourly load energy demand, PV generation output, and energy prices. Day-ahead forecasting data of energy market prices were taken from REE, "Red Eléctrica de España" [29].

Load consumption for a typical residential household was approximated as 11 kWh. A summary of the different parameters used in the residential application case study is presented in Table 4.2.

Load consumption and energy market prices vary significantly from hour to hour and from season to season. There is a close relationship between energy demand and electricity prices. The wholesale energy price increases during summer and winter days compared with a normal day. Furthermore, energy produced from a solar PV system varies from season to season. Hence, it is important to take these differences into consideration when performing a case study. Therefore, input data of energy prices, load consumption, and PV generation for summer, winter, and normal days are used to investigate the effect of seasons on the economic gain. It has been estimated in this case study that a year has 181, 90, and 94 normal, winter, and summer days, respectively. Hourly energy prices and load consumption, along with hourly PV energy generation are shown in Fig. 4.7. It is shown that energy prices are high in the summer compared with other days due to the higher energy demand. Similarly, load consumption and energy prices are higher in winter days than normal days. This demonstrates that an increased energy demand results in higher wholesale energy market prices.

Energy storage revenues

The main objective of the proposed model is the identification of maximum revenues that can be obtained

TABLE 4.2 Specification of Residential Application.	
System Specifications	**Value**
Solar photovoltaic system	3 kWh
Rated capacity of the storage system	11 kWh
Daily operating cycles	1 cycle per day
Storage system lifetime	40 years

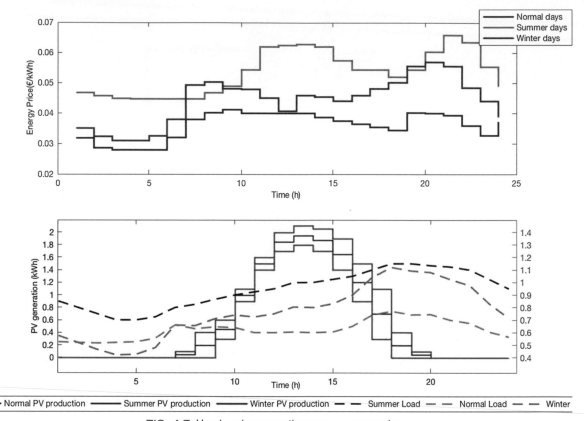

FIG. 4.7 Hourly solar generation versus energy price.

from optimally operating a storage system. The proper scheduling of energy purchase and selling with the grid is necessary to achieve an optimal operation. To increase their economic gain, customers have to buy energy when its price is low and sell it only when its price is high. Fig. 4.8 illustrates the hourly benefits of the first-category scenarios. During normal days, more profit is obtained in the morning (between 8 and 10 a.m.) because of higher energy prices. Inversely, in winter days, more revenues are received in the evening. Concerning summer days, revenues are high between 11 a.m. and 2 p.m. due to high PV production and energy prices. Conversely, even if electric prices are high in the evening, only small economic gain is achieved due to the lower level of storage state and the decrease of PV energy output.

Comparing the benefits delivered from all three scenarios, it can be deduced that the hourly savings are almost the same. The only difference occurs during hours where the prices of energy are identical. In other words, the buying and selling of energy in the

aforementioned hours do not affect the total savings. Therefore, all three scenarios of the first category operate in a similar manner; that is, the obtained maximum revenue of GES is the same in all cases, whether the storage has been coupled to a PV system or enabled/prohibited to provide energy to the demand side. In all situations, the storage system is charged only when energy price is low and discharged when the price increases. Because the benefits of the second category scenario do not vary with time and have been approximated as 550 €/ year, they were not presented along with other scenarios in Fig. 4.9.

After the identification of gravity storage savings, it is necessary to determine the present value of these savings as they would be used for the determination of the system NPV. Fig. 4.10 shows a comparison of present value benefits received from all investigated scenarios. As discussed before, GES in the three scenarios of the first category generates revenues from performing arbitrage service; whereas this system is used in the second group to defer T&D upgrade. The resulted

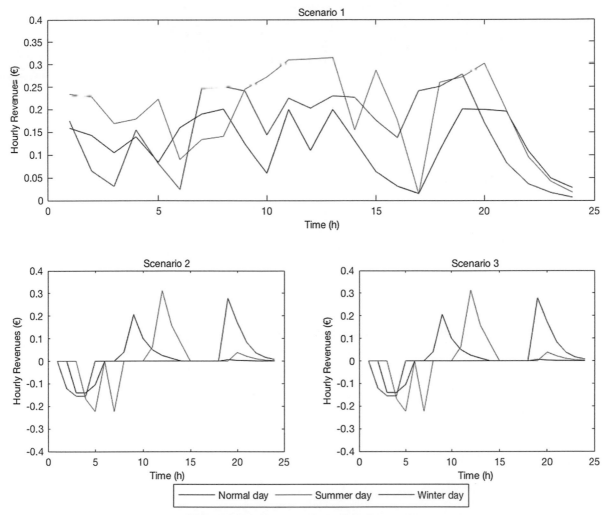

FIG. 4.8 Gravity energy storage hourly revenues.

outcomes demonstrate that the present value of generated benefits is the same for scenario 1, 2, and 3. These are significantly lower than the present value of saving received in the second-category scenario.

Cost of energy storage in residential application

The present value of costs associated with the implementation of GES in the investigated scenarios is determined in this section. This is compared with the cost of other energy storage technologies used in residential applications. Such storage includes lithium-ion and lead-acid batteries with the same scale as GES. These technologies are compared with the use of a number of parameters and costs such as lifetime, capital cost, energy efficiency, and others.

Tables 4.3 and 4.4 summarize the different system characteristics used in this case study. A comparison of all three storage systems has been done using a project life span of 40 years. Because batteries are characterized by a limited life span, replacement costs of these systems are considered an important variable. Based on the investment expected lifetime, the replacement periods of lead-acid and lithium-ion batteries have been calculated as 25 years and 7 years, respectively. As for GES, the system is expected to operate for more than 40 years. Hence, no replacement cost is included for the chosen project lifetime.

In this analysis, the power-to-energy ratio used is 1:2, whereas the usable storage capacity is equal to 11 kWh. The interest rate used for the calculation of the present

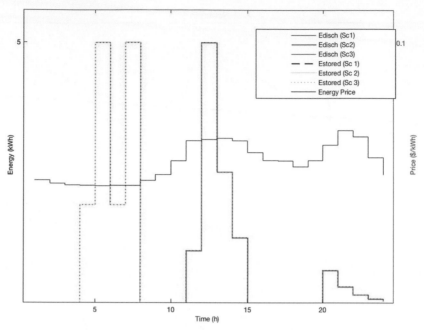

FIG. 4.9 Energy charged and discharged from gravity energy storage.

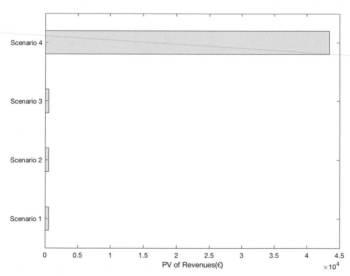

FIG. 4.10 Comparison of scenario's present value.

value of costs is equal to 5%. A number of factors can affect the accuracy of this analysis such as the market conditions, the project location, and the industrial process of the storage technologies. Because lead-acid batteries have a 50% depth of discharge, their installed capacity would be twice the usable capacity. Furthermore, because of their large weight and size, the transportation cost of lead-acid batteries is more expensive than lithium-ion batteries.

Cost per KWh per cycle

Identification of the system cost per cycle is important for understanding energy storage business model. It is calculated by dividing the storage capital cost and

TABLE 4.3
System Characteristics and Costs.

	Installed Capacity	Usable Capacity	Storage Unit Cost	Power Unit Cost	Efficiency	Lifetime	
Lead-acid [8]	22 kWh	11 kWh	150 €/kWh	405 €/kW	75%	500 cycles at 50% DoD	1.5 years
Lithium-ion [8]	11 kWh	11 kWh	700 €/kWh	1350 €/kW	93%	2000 cycles at 100% DoD	5.5 years
Gravity energy storage [9]	11 kWh	11 kWh	31,464 €	600 €/kW	80%	40 years	

TABLE 4.4
System Costs.

	Replacement	Installation	Transportation	Operation and Maintenance	BOP
Lead-acid [8]	30% of equipment cost	200€	28 €/kWh	9 €/kW	90 €/kW
Lithium-ion [8]	50% of equipment cost	200€	10 €/kWh	9 €/kW	90 €/kW
Gravity energy storage [9]	0	200€	-	1.9 €/kW	4 €/kW

transportation and installation costs over the device expected consumption. This is the product of the system number of cycles per year and the number of years the device would be used. Fig. 4.11 shows a comparison of the cost per cycle (€/kWh/Cycle) for the three studied small-scale energy storage systems.

It can be seen that batteries have very high cost per kWh per cycle compared with GES. The reason behind this low cost is because the system has a long life span and does not require any replacement during the studied replacement period. Furthermore, the cost per kWh per cycle of lithium-ion batteries is lower than that of lead-acid even if they require a higher capacity and power unit costs. This is due is to the characteristics and qualities of Li-ion.

Energy storage net present value

To calculate the present value of energy storage cost, Eq. (4.24) is used. The input parameters of these equations include the different energy storage system characteristics summarized in Table 4.3. A comparison of present value of costs of the three studied residential storage technologies is shown in Fig. 4.12. The resulted present value of gravity storage is lower than that of batteries because of the system's long lifetime. Batteries require

frequent replacement, which causes the project investment cost to increase significantly. Furthermore, lead-acid has a higher present value of costs than lithium-ion batteries.

The present value of costs analysis does not demonstrate whether the system is profitable in residential application or not. To investigate this, a NPV analysis should be conducted.

A common methodology to analyze the profitability of a project investment is the NPV. It is the difference between the present value of savings (cash inflows) and the present value of costs (cash outflows) over a specified period of time. In practical terms, it is considered an approach to calculate the project return on investment. You can determine whether the investment is worthwhile by identifying all the expected money you can generate from the project and translating it into today's dollars. A positive NPV indicates a valuable return on investment that is greater than the project costs. Inversely, a negative NPV refers to a return on investments that is worth less than the project costs. NPV of GES for the aforementioned investigated scenarios is illustrated in Fig. 4.13.

The obtained NPV is negative for all the studied scenarios of the first category (scenario 1, 2, and 3),

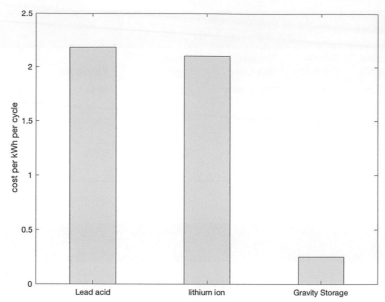

FIG. 4.11 Energy storage systems cost per cycle.

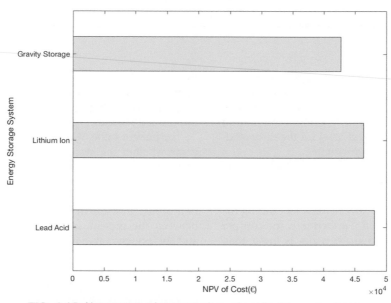

FIG. 4.12 Net present value comparison of residential energy storage.

whereas it is positive for the second-category scenario. It can be deduced that gravity storage is not valuable for small-scale residential application if it performs arbitrage service. However, the system is profitable when used to provide T&D upgrade deferral application. Therefore, depending on the service, it provides

small-scale GES used in residential applications that can obtain a positive NPV and be economically visible.

Case Study for Large-Scale Application

The economic profitability of gravity storage system in large-scale application is evaluated in this section. The

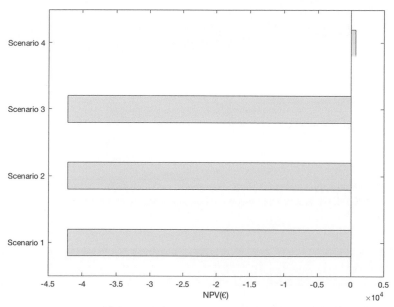

FIG. 4.13 Scenario's net present value.

benefits received from this application include both energy arbitrage and T&D upgrade deferral, which should be determined to calculate the project investment cost. The previously proposed model is used for the identification of the storage maximum saving. The system benefits are the sum of the revenues obtained from performing energy time-shift and T&D upgrade deferral. The system cost in large-scale application has been calculated using GES characteristics and costs summarized in Table 4.5. Fig. 4.14 shows the NPV of the project investment, as well as the present value of the system costs and savings.

A positive NPV of GES demonstrates that the system is profitable for large-scale application. That is, the system is able to generate saving while operating. Therefore, gravity storage is considered economically feasible for large-scale applications.

A payback analysis is considered one of the most common methods to access the economic value of project investment. The payback period of GES has been calculated using Eq. (4.5) and has been found equal to 12 years. Furthermore, the rate of return, which is the inverse of the payback period, is equal to approximately 8%.

To evaluate the competitiveness of gravity storage with other storage systems used for large-scale applications, it is necessary to determine the NPV of these technologies. Such systems used in bulk applications include CAES and PHS. A summary of these systems' characteristics is presented in Table 4.6. The investment period for this simulation has been chosen as 50 years, which corresponds to the lifetime of GES and PHS for large-scale application.

TABLE 4.5 System Costs [9].	
Parameters	**Gravity Energy Storage**
System capacity	20 MWh
Storage power	5 MW
System storage unit cost	1540.32 €/kWh
Power unit cost	600 €/kW
Operation and maintenance	1.9 €/kW
BOP	4 €/kW
Replacement	0
System efficiency	80%
System lifetime	50 years
Other cost including	200 €
• Design	5%
• Acquiring permit	0.5%
• Land	50 €/m²

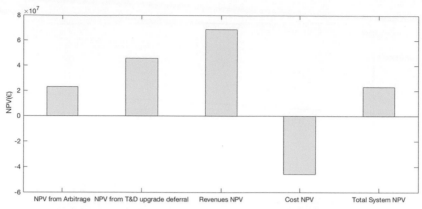

FIG. 4.14 Net present value of large-scale gravity energy storage.

TABLE 4.6
Compressed Air Energy System (CAES) and Pumped Hydrostorage (PHS) System Costs [10].

	Storage Capacity	Power Rating	Storage Cost	Power Cost	Operation and Maintenance Cost	BOP Cost	Rep Cost	Efficiency	Lifetime
CAES	20 MWh	5 MW	45 €/kWh	750 €/kW	1.9 €/kW	4 €/kW	1	80	20–40
PHS	20 MWh	5 MW	93 €/kWh	1860 €/kW	1.9 €/kW	4 €/kW	0	80%	50

Fig. 4.15 shows a comparison of present value of costs and NPV of GES, CAES, and PHS. It is shown that CAES has the lowest present value followed by PHS and GES. Similarly, the NPV is lower for CAES than the two other systems. This is due to its lower capacity and power unit cost compared with its counterparts.

Sensitivity Analysis

The profitability of the system is significantly affected by the different variables used in the performed analysis. That is, a change in one of these variables can impact the system economic viability. To investigate the effect of a change in the input variable on the system profitability, a sensitivity analysis is conducted in this section. The most important risks that can have an impact on the system variables include financial, economic, operation, and political risks.

Economic and financial risks

Investment cost is one of the key determinants of the project profitability. A change of this cost is considered an important risk affecting the system's economic feasibility. The major component of investment cost is the initial capital cost. This later would also have a significant implication and is considered a major economic risk. Fig. 4.15 illustrates the impact of an increase of the system cost on the NPV of the project. It is shown that an 18% increase of the system cost would result in a negative NPV. This means that the system becomes unprofitable in the occurrence of unexpected change in the system costs.

The profitability of the system may also be affected by financial risks. The risk of interest rate variability is investigated, in this study, as it has been used in the calculation of present value of revenues and costs. One of the causes of interest rate volatility is the economic crisis. Therefore, it is necessary to investigate the profitability of the system while taking into consideration the volatility of interest rate. Fig. 4.16 shows also the impact of interest rate on the system NPV. It is shown that the NPV of GES decreases significantly with each percent increase in the interest rate. The system may not be economically viable in countries with high project discount rate. Beyond an 8% increase in the interest rate, the system is not profitable.

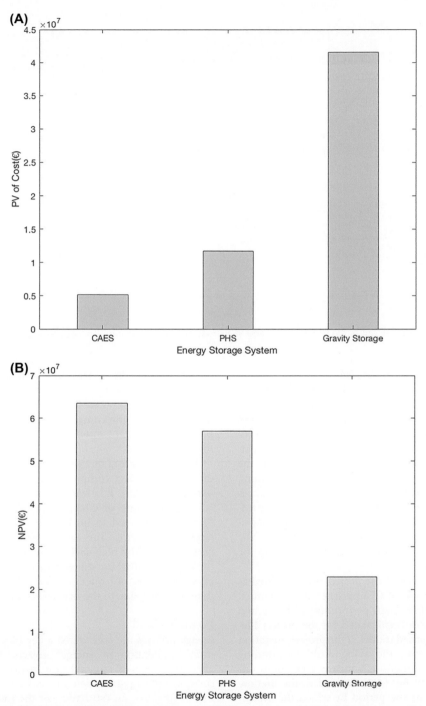

FIG. 4.15 **(A)** Present value of cost and **(B)** net present value of pumped hydrostorage, gravity energy storage, and compressed air energy system.

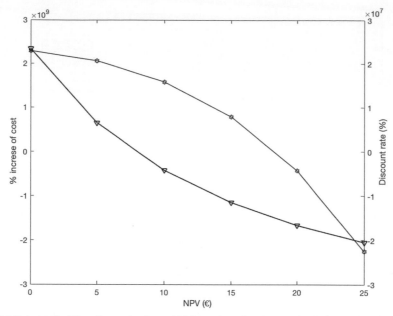

FIG. 4.16 Impact of the discount rate and % increase of system cost on the net present value.

Completion and operation risk

The operation risk of GES may have also an impact on its profitability. This risk could result from a delay in the construction completion or the nonoperation of the system. Such delay should be expected in projects requiring a long construction process. Postponing the completion or the operation of the system which may be due to the occurrence of a fault, would decrease the revenues received. This reduces the NPV obtained and hence affects the system profitability. Therefore, it is important to perform a sensitivity analysis on this risk. Fig. 4.17 illustrates the impact of the operation and construction risks on the system profitability.

The operation delay of the project is determined as follows. An estimation of GES present value of annual savings obtained from both performing arbitrage and T&D deferral services is calculated for the project lifetime using the proposed model. If the project operation is postponed for k years, the saving would start; starting year k of the system lifetime, which is 50 years in our case. Therefore, n in Eq. (4.2) is equal to the lifetime of the system minus the period by which the system has been delayed (k) years. It should be noted that the revenues obtained at year k represents future values of the current year. Therefore, the present value of year k saving should be determined using Eq. (4.1). After taking into consideration the operation delay on the

project savings, the system NPV is then calculated to investigate its impact.

Similar procedures are used for the calculation of the project NPV for the completion delay risk; except the period (n) in Eq. (4.2) which is equal to the system's life span.

It can be deduced that the system remains profitable even if the project completion is delayed for a long period (Fig. 4.16). It is important to mention that the construction time of GES is much less than that of PHS. It is expected to be between 2 and 3 years for large-scale systems. The reason behind this positive NPV is due to the high saving it makes during its long lifetime. Nevertheless, the system profitability may be affected if it is not operated for a long time (more than 30 years) due to the occurrence of a fault.

Political risk

Energy policies are considered a key player in the economic evaluation of energy projects. Political risks affect the revenues received from performing a service such as T&D upgrade deferral and energy time-shift. In this analysis, an estimation of the savings received from T&D upgrade deferral has been done. The real value obtained could be lower than what has been estimated. Therefore, it is crucial to study the effect of a change in the estimated T&D upgrade deferral revenues. Furthermore, it is known that the hourly energy prices

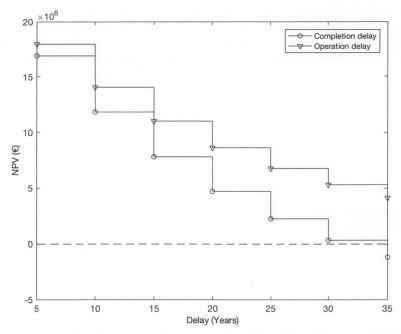

FIG. 4.17 Impact of project construction and operation delay on the system net present value.

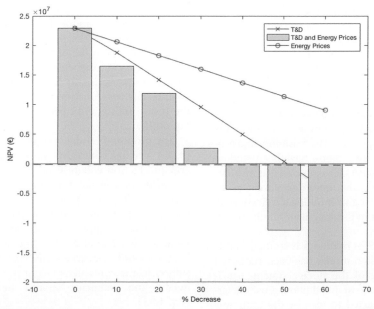

FIG. 4.18 Net present value of gravity energy storage versus % decrease in revenues and energy price.

in the wholesale and retail markets are different from day to day and location to location. Hence, it is important to take into consideration the occurrence of this risk while examining the system profitability.

An investigation of the system economic viability with a decrease of energy prices and T&D upgrade deferral revenues is done in this analysis. Fig. 4.18 shows three scenarios. The first one deals with the effect of a decrease in the hourly energy prices on the system NPV. It is shown that a decrease in energy prices leads to a decrease in the NPV. It should be noted that more than 60% decrease in energy prices does not result in a negative NPV. The second case is about a decrease in the incomes received while performing T&D upgrade deferral. In this case, a 50% decrease results in a negative NPV of the system. The third scenario combines the two aforementioned scenarios. The outcomes of this last scenario demonstrate that the system becomes unprofitable in the occurrence of a 33% decrease in the savings received from the two performed services.

FEASIBILITY OF GRAVITY ENERGY STORAGE IN BUILDINGS

Energy Storage in Buildings

Batteries and thermal storage are mostly used in building application. There exists probably only one PHES integrated in buildings namely the Goudemand resident [11]. This building complex is located in Arras, France. The common area of the apartment building makes use of PVs, wind turbines, battery energy storage, and pumped hydro system. Electricity is supplied to the residence common areas from PV panels and wind turbines. The energy produced from these generation systems is used also to charge the batteries, which supply energy in case the produced energy is less than the demand. The role of PHS comes after the batteries are discharged to a specified depth of discharge; PHS is used to charge these batteries and inject the remaining energy to the local grid. In case the local energy production exceeds the energy demand and batteries are fully charged, energy is stored in PHS by pumping water from the lower reservoir to the upper one. The feasibility and economic viability of such installation have been studied by authors in Ref. [11]. With GES considered as an alternative to PHES storage, the question stands as to whether GES could be used in building application by integrating the container of the system within the building itself.

The feasibility of GES in buildings is studied in this section. An optimal design model has been proposed to investigate the sizing of the storage system, which would minimize the total cost. In addition, a parameterization analysis is conducted to identify the economic profitability of GES in buildings. Competitiveness of this system with other storage technologies used in building application is investigated.

Optimal Design and Sizing

Integrating GES in building requires solving an optimization problem, which would lead to an optimal system size. The height of the storage technology is the same as the building height, whereas its storage capacity is specific to the building needs. The impact of the piston size is considered an important factor in the total system cost. Therefore, a cost minimization problem is defined based on this information to determine the optimal piston size. The cost of the piston is dependent on its mass, which is a combination of the piston height and diameter. The objective of this storage optimal sizing model is to minimize the piston construction cost for a specific energy production while avoiding system failure. The minimization model is solved as a nonlinear program, Eq. (4.29):

$$\min_c f(C) = \frac{1}{4} P_l \rho_p \pi D^2 h_p + P_{cs} \rho_s \left(\frac{1}{4} \pi D^2 + \pi t D h_p \right) \quad (4.29)$$

Subject to

$$\frac{1}{2} (\rho_p - \rho_w) \pi g \mu D^2 h_p (H_c - h_p) - E = 0 \quad (4.30)$$

$$\frac{h_p}{D} > \frac{1}{2} \quad (4.31)$$

$$h_p \leq H_c \quad (4.32)$$

$$D, h_p > 0 \quad (4.33)$$

The aim of this model is to minimize the objective function, which is the piston cost by properly sizing it. Therefore, the value of the height of the piston (h_p) and the piston diameter D (inner diameter of the container) are optimally solved while satisfying all constraints. The cost function f(C) is defined by the cost of the piston filling and casing materials as presented in Eq. (4.29).

The proposed model is constrained by Eqs. (4.30)–(4.33). The first constraint (Eq. 4.30) represents the energy production equation as the system is being sized according to a given storage capacity (E). The second constraint (Eq. 4.31) is used to prevent the piston jamming inside the container. The container's high is provided as an input for the optimization problem. Because the height of the container is known, the piston height should not exceed this; therefore, the model is subject to the container's height limitation. The piston height and diameter should be strictly positive (Eq. 4.33).

A case study is performed to test the effectiveness of the proposed model. A 20 m height building is used as an application of the model with an energy storage capacity of 3.7 kWh. These values were used by authors in Ref. [11] to investigate the feasibility of integrating PHS in buildings. The same conditions and economies of scale are used in this case study to obtain a fair

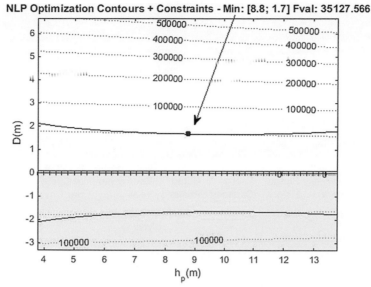

FIG. 4.19 Optimal piston height and diameter.

comparison of the levelized cost of energy (LCOE), which will be discussed in section 4.5.3.

MATLAB OPTI TOOLBOX is used to solve this nonlinear optimization problem. The obtained results illustrated in Fig. 4.19 provide the minimum piston construction cost, which was found equal to 35,127.5 Euros. The optimal piston height and diameter are equal to 8.8 and 1.7 m, respectively.

Parameterization
Cost associated with GES components including piston, return pipe, container, as well as power house equipment are identified and presented in Table 4.6.

Pump/turbine
It is important to properly select the most suitable type of mechanical equipment, which would be used by GES integrated in the building. There are a number of pump/turbine types available. However, these systems should be able to work under the specified site-conditions. Therefore, the selection process should be based generally on the system size, cost, and flow variation. The type of pump/turbine chosen will affect the economics of the system in a building as it would impact both the total cost of the system, its efficiency, as well as its energy production.

The most widely used pump that is suitable for GES system in a building is centrifugal pump. This type of pumps is characterized by its low cost and its ability to operate in diverse working conditions. The estimated

cost of a system with a 20-m pump head and rated power of 1.5 kW is €1.8/W [11]. The installation cost of the system should also be taken into consideration. The efficiency of the pump used is low as small-scale pumps are less efficient and more expensive than their larger-scale systems in terms of cost per power capacity. Hydraulic turbines for small-scale applications requiring low power are not much produced by manufacturers [12]. Therefore, it is more convenient to use pump as turbine in GES building application.

Electronics
Electronic equipment is necessary for the control and proper operation of GES. Sensors are also used to send data to the controller. This system is responsible for switching the pump/turbine and directing the flow by the closing and opening of the electrovalve. The total cost of acquiring and installing sensors, controller, and actuators is estimated to be €4500. The system is automated and programmed at a cost of about €2500.

Return pipe and others
The return pipe used to link the container to the powerhouse could be made of various types of pipes. In this study, rigid PVC pipes are used. The average cost of this type of pipes is €500/m²/m for an area superior to 0.01 m² [11]. Other costs associated with the development of the system include installation, project design, system maintenance, space required. These costs are site dependent. Because GES and PHS

TABLE 4.7
System Costs.

	Piston	Pump/Turbine	Return Pipe	Electronics	Others
Cost (€)	35,127.56	3500	2000	7000	2000

TABLE 4.8
Case Study Parameters.

	Container Height	Container/ Piston Diameter	Piston Height	Energy Capacity	Rated Power	Discharged Energy per Year	Number of Cycles per Day	Discount Rate	Duration of Project
Parameter	20 m	1.7 m	8.8 m	3.7 kWh	1.5 kW	1000 kWh/ year	1	5%	25 years

use similar equipment, the aforementioned costs are approximated to 2000 €.

The main costs associated with the incorporation of GES in buildings include the cost of the piston, the return pipe, pump/turbine, electronics, and other cots. The container structure is considered to already exist as it is part of building. The cost of this is represented by the system space cost included in other costs. Table 4.7 shows a summary of the different costs used for the integration of GES in buildings.

Feasibility Study
Economic viability of gravity energy storage in buildings

The LCOE approach is widely used to value and compare energy storage systems with different characteristics. This method defines a unit cost of electricity generation over the life of the system. It determines the price per energy unit, which balances out the total costs of the system. The LCOE is calculated by dividing the total capital cost of the storage, over the expected energy output while talking into consideration the time-varying value of money.

The LCOE (€/kWh) is obtained using Eq. (4.34):

$$LCOE = \frac{C_C}{E_{out}\left(\frac{(1+i)^n - 1}{i(1+i)^n}\right)} \quad (4.34)$$

where C_c is the cost in year n, E_{out} is the energy output in year n (kWh), i is the real yearly discount rate (%), and n is the project duration in year [13].

Using the costs approximated in the previous section, the LCOE for GES in building was calculated.

The parameters used in this calculation are presented in Table 4.8. The obtained LCOE is equal to 3.69 €/kWh. This value is greatly affected by the piston cost, as it constitutes a major portion of the total capital cost. Therefore, the LCOE for GES in building can be reduced by minimizing this cost further. In addition, the LCOE can be decreased through the reduction in the capital cost or the increase in the system efficiency and output energy.

Competitiveness

It is crucial to investigate the competitiveness of this system with other storage technologies used in this application. This is achieved through a comparison of the calculated LCOE with other energy storage systems used in buildings. Authors in Ref. [11] determined the LCOE of PHS in buildings under the same conditions used in this case study. The used building has a height of 20 m with a reservoir capacity of 80 m³. Two case studies have been performed by these authors and have resulted in an LCOE of 3.59 and 1.66. The value obtained in the first case is close to the LCOE of GES. The second scenario that considers the lower reservoir to already exist has a very low LCOE compared with its counterparts.

A comparison with other energy storage systems used in building applications such as batteries has been also performed. The most widely used batteries in buildings are lead-acid and lithium-ion batteries. The LCOE of these systems calculated under the same conditions has been found equal to €0.78/kW and €0.45/kW, respectively. Fig. 4.20 illustrates a comparison of the resulted LCOE of the studied technologies.

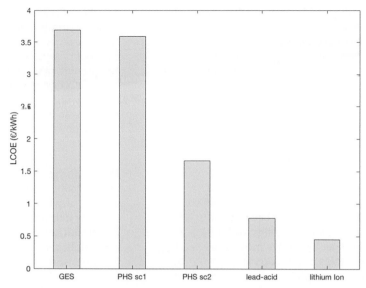

FIG. 4.20 Energy storage levelized cost of energy comparison.

As shown, there is a high difference between the LCOE of batteries and that obtained by PHS or GES. Although these batteries have a short lifetime, they are less expensive in small-scale applications than GES and PHS. Therefore, the cost of constructing GES system needs to be reduced significantly to become competitive with other energy storage in building application.

Cost for incorporating GES in buildings can be reduced through synergies for the system dimensioning and equipment. The installation of the system should be large enough to make use of pump/turbine with higher capacity and efficiency, as well as lower specific cost. Furthermore, it is important to optimize the dimensioning of the pump/turbine as it has a direct impact on the system capital cost and hence significantly affect the LCOE.

CONCLUSION

An important role is being played currently by energy storage in the electric utility system. The use of these systems in the current grid is mainly to store energy generated by large facilities and discharge it based on demand. The current financial benefits of energy storage are mainly received from energy arbitrage, which is about purchasing low-priced energy and selling it when prices are high. In addition to the aforementioned service, other functionalities are provided by energy storage such as black-start, power quality, reliability, and transmission congestion. The main purpose of all

these services is to improve the operational flexibility of the electric power industry. Energy storage is expected to play a significant role in the future electric grid by facilitating and overcoming challenges arising from the increasing penetration of renewable energy systems.

For a residential small-scale application, the combination of energy storage and PV systems is only worthwhile in case specific regulatory and market situations are available to create a favorable and valuable storage investment. This means when the value of energy storage technology is greater than the cost of implementing it. This can occur, for instance, when excess energy generation is used to charge the storage for later consumption; in that situation, end users are required to purchase less energy from the electric grid and hence reduce their energy bills. An economic analysis has been presented in this chapter to investigate the profitability of GES. A mathematical model was developed to determine the maximum revenues obtained from the use of gravity storage in different grid applications. For small-scale application, a number of scenarios have been depicted to represent the different available possibilities of using energy storage in residential application. The outcomes of the performed simulations demonstrate that all scenarios of the first category, which represent storage connected to the electric grid result in a negative NPV. That is, GES is not considered economically viable for residential application when it is used to perform energy time-shift. However, GES system is profitable when it provides T&D upgrade deferral service as it results in a positive NPV. Additionally, for large-scale application, gravity storage has been proven as an economically viable storage solution.

The performed profitability analysis required the use of a number of input variables. A change in these could have an impact on the performed economic viability analysis. To investigate their effects, a sensitivity analysis has been conducted in this work. The project identified risks include economic, financial, operational, completion, and political risks. A positive NPV was obtained for a large range of variation in all the simulated scenarios. However, they were cases resulting in a negative project NPV. For instance, beyond an 18% increase in the system investment costs, the project would not be considered profitable. Additionally, it has been shown that the project has an economic vulnerability with regard to the project interest rate. Yet, a 3% increase in this results in a negative NPV. Furthermore, the outcomes obtained in the simulated political risk, which deals with a decrease in the revenues received from performing storage grid services demonstrate that this has a significant impact on the project economic viability. Finally, operation and construction risks are not considered major risks as the possibility of obtaining a negative NPV is very low compared with other risks. For instance, the project would be considered unprofitable if its completion is delayed for more than 30 years.

Incorporating GES into a building is technically feasible although it is not economically competitive. The performed study shows that gravity storage is an ill-suited solution for building application. Small-scale energy storage systems used in buildings such as batteries demonstrate better LCOE and perspectives compared with GES and PHSs. The high LCOE obtained for GES can be explained by the limited height of the container, which results in a low energy capacity.

NOMENCLATURE

BOP_c	Balance of plant cost (€)
C_c	System capital cost (€)
C_s	Storage capacity cost (€)
d	Replacement period
D	The piston diameter (m)
E	Energy storage capacity (kWh)
E_L	Storage capacity limit (kWh)
E_{max}	Maximum storage capacity (kWh)
E_{out}	Energy output in year n (kWh)
F	Future value
h_p	Height of the piston (m)
Hc	Height of the container (m)
i	Discount rate
j	Number of payments
k	Project operation delay (years)
n'	Number of storage replacement
n	Number of periods (years)
NPV	Net present value (€)
$O\&M_c$	Operation and maintenance costs (€)
P	Present value
P_c	Power costs of the storage system (€)
P_O	Storage power (kW)
P_I	Iron price (€)
P_{cs}	Casing cost €)
$P_E(t)$	Hourly energy prices (€/kWh)
$P(cost)$	Present value of cost (€)
U_{BOP}	Balance of plant unit cost (€/kW)
U_c	Capacity unit costs of the storage (€/kWh)
U_p	Power unit costs of the storage (€/kWh)
$U_{O\&M}$	O&M unit costs of the storage (€/kWh)
Rev_s	Revenues with energy storage
Rev_{ws}	Revenues without energy storage
R_p	Replacement of the storage system's cost (€)
R	Future replacement cost (€)
$S_L(t)$	Storage level (kWh)
$E_{sold}(t)$	Energy sold from the grid (kWh)
$E_{purchased}(t)$	Energy purchased from the grid (kWh)
$E_s^g(t)$	Energy injected in the grid from the storage (kWh)
$E_{PV}^g(t)$	Energy sent to the grid from the PV system (kWh)
$E_g^s(t)$	Energy bought from the grid to charge the storage (kWh)
$E_g^L(t)$	Energy transferred from the grid to the load grid in model Y (kWh)
$E_{Demand}(t)$	Hourly energy demand (kWh)
$E_s^L(t)$	Energy supplied to the load from the storage (kWh)
$E_g^L(t)$	Energy supplied to the load from the grid (kWh)
$E_{PV}^L(t)$	Energy supplied to the load from the PV system (kWh)
$E_{PV}(t)$	Energy generated by the PV system (kWh)
$E_{PV}^s(t)$	Energy generated by the PV system and used to charge the storage (kWh)
$E_g^s(t)$	Energy charged in the storage from the grid (kWh)
$E_{PV}^g(t)$	Energy generated by the PV system and injected to the electric grid (kWh)
$E_{stored}(t)$	Energy charged in the storage (kWh)
$E_{discharged}(t)$	Energy discharged from the storage (kWh)
$E_{sold}'(t)$	Energy sold from the grid without ES (kWh)

$E'_{Purchased}(t)$ Energy purchased from the grid without ES (kWh)

$E'_{Demand}(t)$ Hourly energy demand without ES (kWh)

$E'_{PV}(t)$ Energy generated by the PV system without ES (kWh)

μ System efficiency

δ Storage self-discharge

ρ_p Piston density (kg/m^3)

ρ_w Water density (kg/m^3)

REFERENCES

[1] Huggins R. Energy storage: fundamentals, materials and applications. Springer; 2016.

[2] Kousksou T, Bruel P, Jamil A, El Rhafiki T, Zeraouli Y. Energy storage: applications and challenges. Sol Energy Mater Sol Cell 2014;120:59–80.

[3] Sharma A, Tyagi VV, Chen CR, Buddhi D. Review on thermal energy storage with phase change materials and applications. Renew Sustain Energy Rev 2009;13(2): 318–45. Elsevier.

[4] Energy Storage Association. Applications of energy storage technology. Available from: http://energystorage.org/.

[5] Eyer J. Electric utility transmission and distribution upgrade deferral benefits from modular electricity storage.

Available from: http://prod.sandia.gov/techlib/access-control.cgi/2009/094070.pdf.

[6] Rohit A, Rangnekar S. An overview of energy storage and its importance in Indian renewable energy sector: Part II — energy storage applications, benefits and market potential. Energy Storage 2017;13:447–56.

[7] Walawalkar R, Apt J, Mancini R. Economics of electric energy storage for energy arbitrage and regulation in New York. Energy Policy 2007;35(4):2558–68.

[8] Battery cost. PowerTech System. http://www.powertechsystems.eu/.

[9] Berrada A, Loudiyi K, Zorkani I. Sizing and economic analysis of gravity storage. J Renew Sustain Energy 2016;8. 024101.

[10] Aneke M, Wang M. Energy storage technologies and real life applications — a state of the art review. Appl Energy 2016;179:350–77.

[11] de Oliveira e Silva G, Hendrick P. Pumped hydro energy storage in buildings. Appl Energy 2016;179:1242–50.

[12] Jain S, Patel R. Investigations on pump running in turbine mode: a review of the state-of-the-art. Renew Sustain Energy Rev 2014;30:841–68.

[13] de Oliveira e Silva G, Hendrick P. Lead-acid batteries coupled with photovoltaics for increased electricity self-sufficiency in households. Appl Energy 2016;178: 856–67.

CHAPTER 5

System Performance and Testing

INTRODUCTION

Concerns regarding renewable energy supply security and grid stability are being caused due to the fluctuating nature and the unreliability of renewable energy sources (RES). Therefore, the penetration of the RES into the grid might significantly affect the grid operation at various time scales. To overcome the voltage and frequency issues, regulation frequency and control reserves are needed from seconds to minutes. Load leveling is necessary from minutes to hours at a larger scale [1,2]. These challenges can be dealt with using different methodologies to provide extra flexibility to the energy system. Such methods include electromobility, demand response systems, or flexibility capabilities using conventional generation. Energy storage systems (ESS) are also considered a flexibility solution because of not only the high number of available storage systems but also the numerous services they can offer [3–5].

Although a number of studies have been conducted to analyze the dynamic behavior and the performance of ESS, extra work to investigate their impact on the electric network and the integration of renewable energy systems is critical. A number of studies are about the modeling of isolated and off-grid systems; the outcomes of these analyses have shown that compared with conventional energy systems, the combination of energy storage and renewable energy systems appears to be competitive. Therefore, authors in Ref. [6] examined the performance and ability of ESS to support the use of PV in a residential stand-alone system. The obtained results have shown that grid-independent operation is achieved by the combination of batteries and reversible fuel cells (RFCs), while the use of this latter with supercaps does not meet load demand due to the system's low energy density. Similar studies with an aim to optimally operate and size a PV system coupled with storage technologies such as fuel cells and batteries have been conducted in Ref. [7]. Authors in Ref. [8] examined a dynamic model and developed a control system for a PV/battery combination using MATLAB; the outcomes of this analysis demonstrate that the control strategy can regulate frequency and voltage regardless of the load variability and PV

generation uncertainty. As for grid-connected systems, many interesting studies have been performed by several researchers. For example, microgrid applications and power systems with a high integration of renewable energy resources can be found in Refs. [9–15].

Modeling the dynamic of ESS have been tackled by several works. These systems may be categorized based on the form of energy stored which can be chemical, electrical, thermal, or mechanical energy storage. As for the first category, two mathematical models were developed by authors in Ref. [16] for calculating the dynamic current/voltage parameters of (PEMFC) fuel cells. The dynamics of a hybrid system comprised of a battery and hydrogen storage connected to a PV system were modeled by Douglas [17]. The objective of this analysis is to study the performance of this configuration. The main findings of this study have shown that battery can be operated for longer time and can meet the load with low power as it can be used for low power loads; whereas fuel cell can satisfy high power loads as it can operate for shorter periods. The fuel cell dynamic response was partial due to the reactants flow conditions nearby to the electrodes, which results in a rise in the internal resistances. A high-pressure polymer electrolyte membrane electrolyzer was modeled by Yigit et al. using Simulink [18]. The purpose of the analysis is to study the behavior of the system and approximate its losses at drivers' operating conditions. The outcomes of this study demonstrate that the efficiency of the system decrease considerably at high current density.

As for thermal energy storage, a review has been published by authors in Ref. [19] concerning the different approaches used for modeling energy storage in buildings application. Their study emphasizes on developing and modeling the system control strategies. The transient behavior of refrigeration and the phase change material energy storage has been modeled and simulated by Wu et al. [20]. This work makes use of differential equations to represent the storage system. Equivalent resistance circuits with lumped parameters and electric capacitance have been used to model the system. The obtained simulation results were quite similar to the experimental ones [20]. Modeling the

Gravity Energy Storage. https://doi.org/10.1016/B978-0-12-816717-5.00005-0

dynamics of power and high-temperature heat storage system has been carried out in Ref. [21]. Underground thermal energy storage was used in this study. Concerning electric energy storage, a Simulink model for high-temperature SMES systems was developed in Ref. [22]. The system control strategy has been tested in this study by the use of a simulation.

ESS that are widely developed for large-scale applications include compressed air energy storage (CAES) and pumped hydroenergy storage (PHES); the installed capacity of this latter represents 98% of the total installed storage capacity worldwide [23]. To analyze the behavior of mechanical ESS, a number of dynamic models have been proposed. Authors in Ref. [24] modeled the performance and ability of CAES system to support the integration of wind energy. The obtained results have shown that due to their high-energy density, hydrogen-based systems are able to store more power than air-based system. A dynamic model with a control strategy has been developed by Saadat et al. in Ref. [25] to simulate a novel CAES connected to a wind farm. As for isobaric CAES, authors in Ref. [26] developed a model to investigate the system transient state. The aim of this study is to assess the system response time and ability of responding to load demand.

Pumped hydrostorage is typically modeled in a similar manner as hydroelectric power plant. There exist in literature a number of published models dealing with the dynamic response of hydropower systems. In addition, the different parts of this latter, which include the penstock, the hydroturbine, the generator, and the governor, have been modeled in Ref. [27]. The performance of the hydroturbine is influenced by the flow properties, such as inertia, compressibility, and elasticity of the penstock walls [28]. The hydroturbine system has been modeled considering the linearity and nonlinearity approach, as well as the effect of the elastic and nonelastic water column. Authors in Refs. [29,30] proposed the conduit system models. The nonelastic water column effect has been considered by Ref. [31]. Authors in different research papers have used a number of assumptions to represent the output of the hydroturbine model. For example, in Ref. [32], authors have assumed that the output of the hydroturbine is proportional to the product of the head and the fluid flow rate. The first-order Taylor formula was used by authors in Ref. [33] to model the hydroturbine system. The development of the hydroturbine and governor model is a difficult task as it combines the mechanical, electric, and hydrodynamics [34–36]. The nonlinear dynamic modeling of this latter system with surge tank has

been carried by authors in Ref. [36]. As for the load and the generator system, a number of models, which include first, second, and third order have been proposed by different researches [37–39]. Some authors have modeled the turbine system using a nonlinear methodology because large variation in power output may affect the power system performance [40]. On the other hand, small variations were taken into consideration by authors in Refs. [34,36,41–49] to model the hydroturbine governing system. In Ref. [50], the process of load reinjection transient is considered. This analysis provides a detailed study about the nonlinear dynamic behavior of the hydroturbine governing system.

The organization of this chapter is as follows. Section 2 discusses the governing equations of the system dynamic models. A case study simulation with an analysis of the system behavior is presented in Section 2. Section 3 provides models for the system hydraulic components. Section 5 analyzes the simulation results and compares them with the experimental ones. A sensitivity analysis is performed in Section 4 to examine the impact of critical parameters on the simulation obtained results. Section 5 closes the chapter.

DYNAMIC MODELING OF GRAVITY ENERGY STORAGE

The objective of this mathematical model is to investigate the response of gravity energy storage (GES) coupled to a photovoltaic (PV) farm. The energy demand of the residential load has to be met by the hybrid energy system connected to the electric grid. The proposed model is developed using governing physics equations. These latter are based on the electronics, mechanics, and hydraulics theories. Matlab/Simulink has been used as a simulation tool to assess the hybrid system performance.

Discharging Phase of Gravity Energy Storage

Over the past few years, a number of hydraulic models have been proposed. This work makes use of a basic hydraulic model to describe gravity storage discharging mode [51]. Investigating the dynamic response of the mechanical components of the system is the main objective of this study. This section introduces the different equations used by the developed model along with some simplifying hypotheses. The input of this submodel includes the system reference power; whereas its output consists of the system-produced power. The model of the hydroturbine is represented by the valve opening dynamics and the interference between the turbine blades and the fluid. The submodel is designed

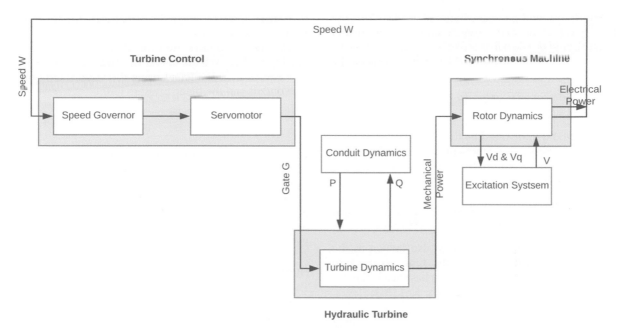

FIG. 5.1 Hydraulic turbine modules.

using blocks available in SimPowerSystems. The inter-action between the hydraulic turbine constituents is illustrated in Fig. 5.1.

The hydraulic turbine block generates mechanical power, which is used to rotate a synchronous generator. This latter produces electrical power. The excitation system block supplies an excitation voltage to the synchronous generator. The proportional, integral, and derivation (PID) controllers are used to regulate both the excitation voltage and the mechanical power of the turbine. The output power produced by generator is fed to a power transformer, which in turn supplies a transmission line. The energy demand is modeled using a dynamic load. To simulate GES discharging phase, the aforementioned modules are modeled in an interconnected and dependent network.

Hydraulic turbine

To represent the hydraulic turbine models, it is commonly assumed that the hydraulic resistance is negligible, water is inelastic and incompressible. In addition, the velocity of water varies as a function of system pressure head and gate opening. It is also assumed that the turbine produced power is proportional to the product of the square root of the system head and the flow rate.

The governing equations relating the power of the turbine, the water motion, and the turbine inlet are used to determine the turbine and the return pipe characteristics [52]. The turbine is modeled using the system hydraulic and mechanical characteristics.

The volume of water entering the turbine is calibrated by varying the opening of the gate, which also regulates the turbine power output. The relationship between the head across the hydroturbine and its flow rate is represented by Eq. (5.1). This is expressed as a function of the valve characteristics [51].

$$h = \frac{Q^2}{G^2} \tag{5.1}$$

G represents the position of the gate which is between 0 and 1. When the gate is fully opened, G is equal to 1. The head across the turbine is h and Q id the flow rate.

The net force applied on the water is used to determine the water flow rate. It is formulated using the rate of change of momentum at steady state conditions, and the pressure head at the return pipe as shown in Eq. (5.2).

$$\begin{cases} F_{net} = \rho L \dfrac{dQ}{dt} \\ F_{net} = \rho(H_S - H_l - H)Ag \end{cases} \tag{5.2}$$

where L represents the water conduit length and ρ the water density. H_s is the static head at the turbine, H the head at turbine gate, H_L are the head losses, A the conduit area, and g the gravitational acceleration. The head loss (H_l) due to the water and the conduit friction effect should be taken into account as well. Eq. (5.3) describes the volumetric flow, which is derived from the combination of two aforementioned equations.

$$\frac{dQ}{dt} = \frac{(H_S - H_l - H)Ag}{L} \quad (5.3)$$

Eq. (5.3) is normalized by the use of h_{base} and q_{base}. Here, the base head is represented by the static head, while the turbine flow rate with fully opened gates denotes the base flow rate. Eq. (5.4) is formulated as:

$$\frac{dq}{dt} = \frac{(1 - h_l - h)Agh_{base}}{Lq_{base}} = \frac{(1 - h_l - h)}{T_W} \quad (5.4)$$

The water time constant (T_w), also known as the water stating time, represents the time necessary to reach the base flow rate with a given base head. The water time constant is expressed as (Eq. 5.5) [51]:

$$T_W = \frac{Lq_{base}}{Agh_{base}} \quad (5.5)$$

The generator shaft is driven by the mechanical power (P_m) provided by the Francis turbine. P_m is dependent on the water flow rate and the system pressure. This latter can be represented by a nonlinear Eq. (5.6) [53–55].

$$P_m = \eta q \rho g h \quad (5.6)$$

The mechanical power produced by the turbine is related to the pressure head and the flow rate. The power is also related to the efficiency η and is represented by no-load flow (q_{nl}).

$$P_m = h(q - q_{nl}) \quad (5.7)$$

Eq. (5.7) can be rewritten as Eq. (5.8) by taking into consideration a different generator per-unit system.

$$P_m = A_t h(q - q_{nl}) \quad (5.8)$$

The difference in per units is represented by the factor A_t, which is formulated by Eq. (5.9).

$$A_t = \frac{Turbine\_Power(MW)}{generator\_MVA\_rating} \frac{1}{h_r(q_r - q_{nl})} \quad (5.9)$$

It is also important to consider the damping effect, which depends on the opening position of the gate. Using the previously mentioned equations, the turbine power is expressed as Eq. (5.10) [51].

$$P_m = A_t h(q - q_{nl}) - D_c G \Delta \omega \quad (5.10)$$

Here D_c and ω represent the damping coefficient and the rotor speed, respectively.

The classical transfer function shown in Eq. (5.11) can be used as well to model the hydraulic turbine. This equation is developed using the water inertia in the conduit and the blade pressure [56]:

$$W_H(S) = \frac{\Delta \overline{Pm}}{\Delta G} = \frac{1 - T_W s}{1 + 0.5 T_W s} \quad (5.11)$$

where $\Delta \overline{P_m}$, and ΔG are the change in power output and gate position, respectively. The equation shows that the hydroturbine response as a function of the mechanical power production over the variation in the gate position. The nonlinear model of the turbine developed using the presented governing equations is shown in Fig. 5.2 [51].

Hydraulic governor
The turbine governor is one of the most essential parts of a hydroplant. The main function of the governor is to monitor the generator speed variations with an aim to ensure a constant frequency. The system components include gate servomotor, relay valve, speed sensing, dashpot, permanent droop feedback, and computing functions [57].

The speed of the rotor is compared with the reference speed using the controller, which modifies the speed with the use of permanent droop compensation. To avoid fast variation of this latter, a transient droop compensation is established during a change in the position of the gate. The servo motor input directs the valve according to the signal created by the variation in the rotor speed. A constant generated frequency is maintained using the valve that monitors the flow rate. Thus, the floating levers system transmits signals from the mechanical motion to the pilot valve [58]. The transfer function of the relay valve and the gate servomotor is as follows [51]:

$$\frac{\Delta G}{b} = \frac{K_2}{s} \quad (5.12)$$

The transfer function of the pilot valve and the pilot servo is [51]

$$\frac{b}{a} = \frac{K_1}{1 + T_p s} \quad (5.13)$$

where K_2 is obtained from the feedback lever ratio and T_p is identified from K_2 and port areas of the pilot valve.

The combination of Eqs. (5.12) and (5.13) gives

$$\frac{\Delta G}{a} = \frac{K_1 K_2}{(T_p s + 1)s} = \frac{1}{T_g} \frac{1}{(T_p s + 1)s} \quad (5.14)$$

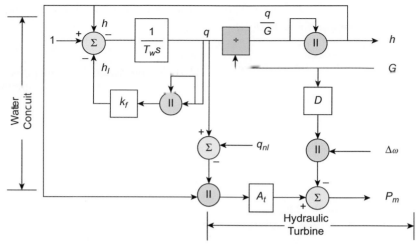

FIG. 5.2 Nonlinear model of the turbine.

The transfer function of the dashpot is given by Ref. [51]:

$$\frac{c}{\Delta G} = R_T \frac{T_R s}{1 + T_R s} \tag{5.15}$$

The temporary drop (R_T) is obtained from the lever ratio, whereas the reset time T_R is obtained by using the valve setting.

Old speed governors work with both hydraulic and mechanical systems. The mechanical components perform functions related to speed sensing, droop feedback, and computing, whereas the hydraulic components perform functions requiring higher power. A dashpot delivers the transient droop compensation. To disable the dashpot, a bypass arrangement is used. On the contrary, the new hydraulic governors use electrohydraulic systems. These systems function as mechanical-hydraulic governors. The electric governor dynamic aspects are adjusted to match the features of the mechanical-hydraulic governors. In this situation, measured tasks such as permanent droop and speed sensing are performed electrically. The electric components provide flexibility and better performance with respect to dead bands and time lags. PID controllers, including the proportional (P), integral (I), and derivation (D) elements, are used in these new systems. To achieve higher respond speeds, the algorithm by which the PID delivers transient gain increase and decrease. For plants characterized with a starting time higher than 3 s, the derivation action of the controller is more suitable. The value of the P, I, and D gains are typically equal to 3, 0.7, and 0.5, respectively. To prevent excessive oscillations and system instability, the derivative gain is set to zero. In latter situation, the controller

turns to PI governor and is much like the mechanical-hydraulic governor. The PI gains can be selected according to the specified temporary droop and reset time [59].

There are a number of functionalities that can be offered by governors with PID control. These include the control of load and the turbine speed, the offering of minimum dead band, the provision of a favorable dynamic response, as well as the establishment of normal conditions.

Synchronous generator and excitation system
The block of the synchronous machine used in this simulation can be operated as a generator or a motor. The sign of the mechanical power is an indicator to the operation mode of the system. To model the electric components of the generator, a sixth order space is used. The circuit model is represented by the rotor reference frame. The model takes into account the inertia of the turbine/generator components, the stator, as well as the dynamics of the damper windings and the field. The system makes use of the excitation voltage and the turbine mechanical power as inputs; while the reactive power and active power are used as outputs. Power plant documentations are used to obtain the values of the generator parameters [60].

The excitation system block supplies an excitation voltage to the synchronous generator. The PID controllers is responsible for regulating the excitation voltage. Eq. (5.16) represents the excitation system transfer function. This is given in terms of the exciter voltage and the regulator output [61].

$$\frac{V_{fd}}{e_f} = \frac{1}{K_e + sT_e} \tag{5.16}$$

where T_e and K_e are the time constant and feedback gain, respectively.

Charging Phase of Gravity Energy Storage

The excess power is stored in GES as potential energy in the container. The pump is used to lift the piston and store energy. Its operation relies in three important parameters, which include the water velocity, flow rate, and the head pressure. The pump operation is expressed by Eq. (5.17).

$$f(H, Q, V) = 0 \qquad (5.17)$$

To solve this equation, one of the three parameters, is considered as a constant. The flowrate is given by Eq. (5.18):

$$Q_v = \frac{V}{t} = vA \qquad (5.18)$$

Here Q_v is the volumetric flow rate(m³/s); V is the water volume (m³); v is the velocity of water in the pipe (m/s); and A is the conduit area (m²). The discharge pressure of the pump is formulated as Eq. (5.19).

$$H_d = \frac{P_d}{\rho g} \qquad (5.19)$$

where H_d, P_d are the discharge head and pressure, respectively. The pressure of the pump at suction is expressed by Eq. (5.20):

$$H_S = \frac{|P_S|}{\rho g} \qquad (5.20)$$

where H_S, and P_S is the suction head and pressure, respectively.

The sum of the suction and the discharge heads results in the pump total head. This latter is formulated as Eq. (5.21):

$$H_T = H_d + H_S = \frac{P_d + |P_S|}{\rho g} = \frac{P_d - P_S}{\rho g} \qquad (5.21)$$

Using Eqs. (5.20) and (5.21), the pressure difference is found by

$$\Delta P_P = P_d - P_S = \rho g H_T \qquad (5.22)$$

The useful and consumed power of the pump are

$$\begin{cases} P_{pu} = \Delta P_P Q_v = Q_m g H_T \\ P_{Cp} = g \dfrac{Q_v \rho H_T}{\eta_P} \end{cases} \qquad (5.23)$$

The efficiency of the pump is given by Eq. (5.24):

$$\eta_P = \frac{P_U}{P_{Cp}} \qquad (5.24)$$

Here P_U and P_{Cp} are the useful power and the electric power consumed by the pump, respectively.

Energy Management Model

The flow of energy in the storage system is controlled by an energy management system. This work deals with two scenarios consisting of the storage and the discharge of energy. GES stores the surpass energy, which is the difference between the load demand and the energy generated by the PV system, when this latter is at a higher quantity. The stored energy is released when the energy demand is smaller than the energy produced by the PV. In this case, the energy released from the storage is equal to the difference between the PV-produced energy and the load. The storage system rated power and capacity limits are considered by the model. That is, the discharged energy at a certain time should not be greater than the energy level in the storage system at that time, and the energy charged should be lower than or equal to the system capacity. Figs. 5.3–5.24 demonstrate the system discharging and storing algorithms.

The model has to verify the storage system state before starting both the storing and discharging processes of the storage system. A Simulink block diagram illustrating the storage state is shown in Fig. 5.5.

Eq. (5.25) presents the storage state at time t:

$$S(t) = (1 - \delta)S(t - 1) - E_D(t) + (E_S(t)\eta) \qquad (5.25)$$

The system storage level is determined form the energy stored at time (t), the energy left in the storage at time $(t-1)$, and the energy discharged from the system at time (t). System losses should also be taken into account; these include the system self-discharge rate (δ) and round-trip efficiency (η). The storage level should not be greater than the storage capacity and must be positive.

Photovoltaic and Grid Models

Eqs. (5.26)–(5.28) were used to model the PV system. A solar array consisting of n modules has a maximum rated nominal power equal to Ref. [62]:

$$P_p = nW_p \qquad (5.26)$$

At standard test condition (STC) and with 1 kW/m² radiation, the system has a peak nominal power (P_p);

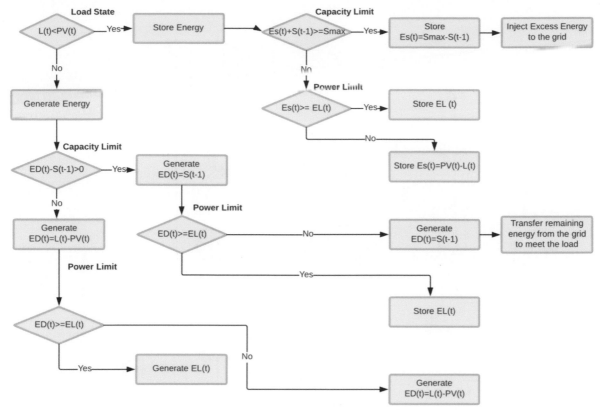

FIG. 5.3 Energy management system flowchart.

The module maximum power is (W_p). The average energy generated per year by the PV plant is expressed as follows:

$$E_p = \frac{S_r P_p}{I_{stc}} \eta \qquad (5.27)$$

Here S_r represents the solar radiation for a specific region. This latter relies on the weather conditions and the period of the year; I_{stc} is the solar irradiance at STC; and η is the solar cell efficiency. The average generated energy is found using Eq. (5.28) while taking into consideration the system orientation and inclination [62]:

$$E = E_p \xi \qquad (5.28)$$

The system efficiency takes into consideration the temperature derating factor. This is included in the overall efficiency of system (75%), and is expressed in Eq. (5.29) [62].

$$\eta_t = 1 \quad [\gamma \times (T_c - T_{stc})] \qquad (5.29)$$

where γ represents the power temperature coefficient. For crystalline silicon, this coefficient is typically estimated as 0.005 [62].

A dynamic load block is used to model the demand load in Simulink. This load is met through the provision of energy from both the electric grid and the hybrid renewable plant. The role of the grid is to optimally dispatch energy between the different elements of the modeled network. The grid model equation is represented by:

$$P_{grid}(t) = P_L(t) - P_D(t) - P_{PV}(t) + P_S(t) \qquad (5.30)$$

where $P_{grid}(t)$ is the power exchanged between the electric grid to the residential load; $P_L(t)$ is the load demand; $P_D(t)$ is the storage discharged power; $P_S(t)$ is the system stored power; and $P_{PV}(t)$ is the PV output power.

Model and Simulation Analysis

To investigate the effectiveness of the model, a simulation is performed on a hybrid solar power plant connected to an ESS. This system is linked to the grid.

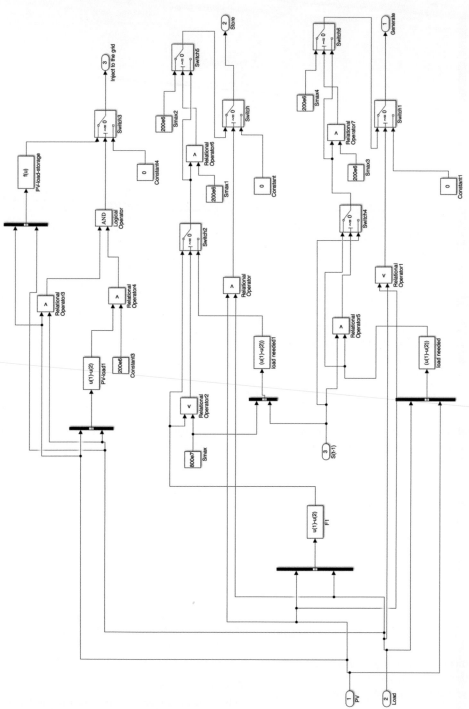

FIG. 5.4 Simulink model of the energy management system.

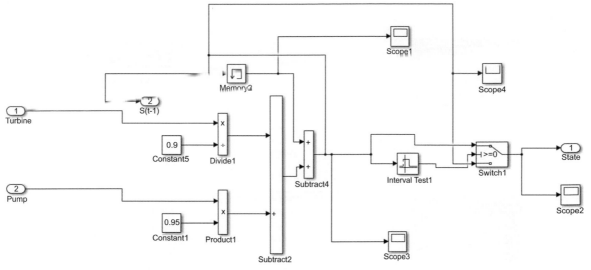

FIG. 5.5 Storage state.

Simulink Submodels

GES submodel in charge of the production of energy is illustrated in Fig. 5.7. It is made up of a number of block diagrams connected to each other, which include a hydroturbine governor, a synchronous generator, three-phase RLC load, and an excitation system.

Power is produced by the hydroturbine block (HTG) and is supplied to the generator. The generator shaft operates at synchronous speed using the mechanical power output to generate electric power. The electromechanical torque and the speed and created by this latter are multiplexed at the terminal "m"; then the output signals are demultiplexed by a bus selector from the same terminal.

The demultiplexed signals include the active power output, the rotor speed, and the stator voltages V_q and V_d as shown in this model. The rotor speed and output active power are supplied to the hydroturbine governor, whereas the stator voltages V_d and V_q are used as feedback signals by the excitation system, which has the responsibility of delivering excitation to the generator.

The actual speed of the generator is compared with the reference speed. The speed error is the difference between the two aforementioned speeds and is used by the PID controller. Reducing the speed error increases the stability of the HTG; this is achieved by accurately selecting suitable constants for the PID controller. The system is stabilized by the output signal of the governor; this latter is used by the servo system of the HTG as input to switch the gate opening of the

Real input data are used in this case study. The main goals of this simulation are to investigate the functioning of the hybrid power plant with regard to responding to residential load; to analyze the output production of the solar plant and the storage; and to determine the power injected to the electric grid. The block representing the residential energy consumption is linked to the solar farm, the energy storage and the grid blocks. To meet the demand, energy output of the PV and the storage systems, as well as the energy transferred from the grid are supplied to the load block. The model is very coupled as presented by the system block diagram (see Fig. 5.3). The following subsection discusses the individual submodels.

The grid, the solar system, and the storage discharging block are all linked to the residential load block. To match the energy demand, the electricity produced by the aforementioned systems along with the energy supplied from the grid are delivered to the residential block. Energy generated by the PV plant and the storage system flow rate at the valve are inserted as input into the storage pump mode block. The power generated by this block is provided to the energy storage block. This latter is responsible for the control of energy being charged and discharged from the storage system. The system characteristics, such as the pressure used to calculate the flow rate, can be determined using the container/piston assembly. This assembly is connected to the valve, which sends the flow rate value to the charging mode block. As shown in Fig. 5.6, the model is highly coupled.

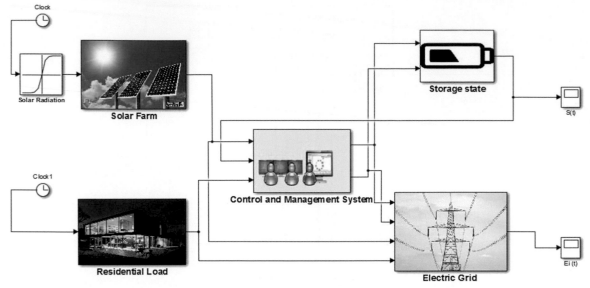

FIG. 5.6 Case study model.

wicket. If the generator speed is greater than the reference speed, the servo system decreases the gate opening as a result of a signal sent by the PID controller. This decrease will induce the reduction of the flow rate as well as the mechanical power produced by the hydroturbine. Hence, the generator speed decreases, and start rotating at synchronous speed. GES charging phase model is illustrated in Fig. 5.8.

Case Study

The model uses the hourly solar radiation and energy demand as inputs; these latter were obtained from Refs. [63,64]. Fig. 5.9A illustrates the hourly solar radiation. The power generated by the PV system is calculated based on the system efficiency and solar radiation. Eq. (5.27) was used to approximate the solar plant energy output. For this case study, a 918 MWh PV power plant with an efficiency of around 75% is modeled. The PV-produced power output is illustrated in Fig. 5.9B. The power generated by the PV system has nearly a bell-shaped curve. A gradual increase is observed in the morning; reaching the maximum production output at 12 a.m.; and decreasing progressively in the afternoon. The energy demand is shown in Fig. 5.10. It varies during the day, reaching its maximum at peak periods and decreasing in the evening.

A 200-MVA synchronous machine is used for this case study. The terminal voltage of the synchronous machine is set at 1pu and is controlled by the excitation system. Table 5.1 [65] shows the model parameters used as inputs for the excitation system, the synchronous machine, and the turbine.

Simulation Results

As previously mentioned, the PV system, the energy storage, and the electric grid are all used to meet the energy demand. Fig. 5.10 shows the energy generation demand curve; a comparison between Figs. 5.10 and 5.11 shows that the demand is matched by a single system or a combination of these systems. The participation of each system is shown in Fig. 5.11. The grid supplies electricity at night when energy is not being dispatched neither from the PV system nor the storage (low storage states). Power produced by the PV system between 5 and 7 a.m. is low; this amount of energy is supplied to the residential energy demand. During this period, power is also transferred between the residential load and the grid. The contribution of both the grid and the PV system is essential to meet the demand. Power is met by only the solar farm from 7 a.m. until 3 p.m. At that time, excess energy produced by the PV system is used to charge the storage. This latter reaches its peak capacity at approximately 3 p.m (see Fig. 5.11).

When the energy generated by the solar farm is small, the contribution of ESS is essential. This occurs mainly between 3 p.m. and 5 p.m. Hence, both the storage and the PV produce energy. At 6 p.m., the load is supplied power from energy storage. The storage system state is shown in Fig. 5.12. The charging of this system

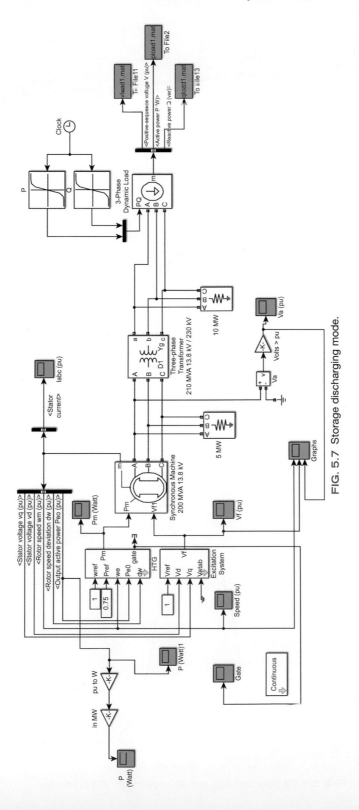

FIG. 5.7 Storage discharging mode.

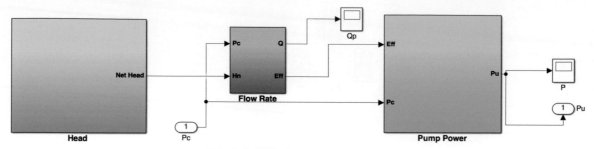

FIG. 5.8 GES charging phase model.

(A)

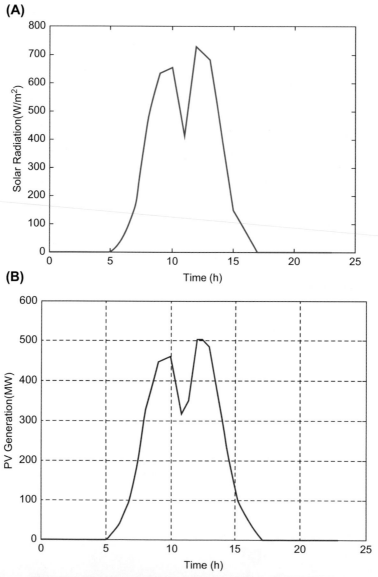

FIG. 5.9 **(A)** Solar radiation and **(B)** photovoltaic (PV) production.

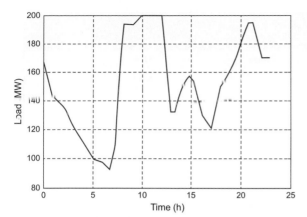

FIG. 5.10 Hourly energy demand.

takes place at 6 a.m. as the solar farm begin producing electricity. The system discharges energy during periods of high energy demand and when the PV farm is not generating electricity. Fig. 5.13 shows the storage charging and discharging. The energy used by the pump system is represented in green, whereas the useful energy of the storage system is represented in red. The energy efficiency of the system explains the disparity between the two values. The energy discharged by the storage system is represented by the blue color.

The energy exchange with the grid is shown in Fig. 5.14 Energy is being transferred from the electric grid in the evening when the solar farm is not generating power and when the storage system is empty. Small amount of power is transferred back to the electric network; occurring at 10 and 12 a.m. At that time, the PV system power output reaches its peak; an amount of this generated energy is supplied to the load; stored in the storage system; and fed to the electric network.

Owing to water inertia, hydroturbines show a peculiar response. A shift of the gate position causes a variation of the hydraulic turbine-generated power. The position of the gate is changed due to a control signal emitted by the governor. The generator that is responsible for generating electric power output is driven by the hydroturbine (see Fig. 5.15). Eq. (5.10) shows the effect of power output variation on the gate position. Fig. 5.16 shows the response of the storage system as the storage switches from standby mode to discharging phase. As shown in this figure, the power produced by the turbine initially is equivalent to half the time Tw needed for the gate to open promptly due to water inertia [56]. The power output appears to be quickly decreasing between 1 and 1.335 s because of an initial

step rise in gate position. This is caused by the negative sign in the numerator of Eq. (5.10) [59]. As the system flow rates gets larger, the power output starts to increase as well.

In contrast, the gate should be closed if the storage system is switching from the discharging operation mode to standby mode with the objective of reducing the power output. In this situation, the flow rate does not vary instantly, which induces a first rise of the flow velocity. This also means that there is a brief first rise of the power output. The latter will decrease after a decrease in the flowrate [59].

The mechanical components used by gravity storage are similar to those used by hydropower plants. These equipment are characterized by a short response time, which make this technology capable of ensuring grid stability and rapidly following variations in energy demand and supply.

Different loading conditions were simulated to examine the response behavior of GES. Figs. 5.17B and 5.18 show the characteristics of the synchronous generator active power. A steady state value of 0.6 pu is shown by this curve; this latter can be converted to the actual load in watts. When the generator starts operating, it results in a steady state. The power characteristics experience undershoots and overshoots because of the few oscillations used to achieve a stable operating point on power.

The voltage (V_a) and the generator stator currents are shown in Fig. 5.17A. The stator phase voltage is the terminal voltage (phase A). It consists of three-phase voltages offset by 120 deg. The curve follows a sinusoidal pattern and shows the steady state characteristics of the voltage. The phase-voltage magnitude, which is equivalent to the generator rated voltage output, is equal to 1 pu. Load variation does not influence the terminal voltage as the latter remains constant.

The stator current is shown in Fig. 5.17B. It appears to be slightly varying due to a load change at 960 min, which is equivalent to 4 p.m. The three-phase current envelop presents an underdamped response when the load is changed. After a period of time, steady state characteristics are achieved. The same characteristics are shown in Fig. 5.17D from a closer view.

Fig. 5.17C shows the excitation voltage (V_f) curve as a function of time. The system operates in steady-state situation before changing the load. An increase of the excitation voltage is shown as the load increases. Similarly, when the resistive load decreases, the excitation voltage also decreases before returning to the steady state condition.

TABLE 5.1
Equipment Parameters.

Component	Parameter	Symbol	Value	Unit
Synchronous machine	Mechanical power	P_m	Pm	MW
	Rotor type	—	Salient-pole	—
	Nominal power		200	MVA
	Line-to-line voltage	V_n	13800	Vms
	Frequency	f_n	60	Hz
	Stator	R_s	2.8544e-3	pu
	Time constants	[Td' Td'' Tqo'']	[1.01, 0.053, 0.1]	s
	Reactances	[Xd Xd' Xd'' Xq Xq'' Xl]	[1.305, 0.296, 0.252, 0.474, 0.243, 0.18]	pu
	Inertia coefficient, friction factor, pole pairs	[H(s) F(pu) p]	[3.2 0 2]	-
	Initial conditions	[dw, th, ia,ib,ic, pha,phb,phc, Vf]	[0—94.2826 0.750185 0.750185 0.750185 -24.943 -144.943 95.057 1.29071]]:	[dw(%) th(deg) ia,ib,ic(pu) pha,phb,phc(deg) Vf(pu)]:
Hydroturbine	Turbine	G_{max}	0.01	—
		G_{min}	0.97518	—
		beta	0	—
		T_w	2.67	s
	Servomotor	Ka	10/3	—
		ta	0.07	s
	PID	Rp	0.05	—
		Kp	1.163	—
		Ki	0.105	—
		Kd	0	—
		td	0.01	s
Excitation system	Transient time constants	Tc, T_B	0	—
	Regulator gain	Ka	300	—
	Regulator time	Ta	0.001	s
	Exciter gain	Ke	0	
	Exciter time constant	Te	0	s
	Damper gain	Kf	0.001	
	Damper time constant	Tf	0.1	s

Fig. 5.18 shows the rotor speed characteristics. The synchronous speed steady state value is equal to 1.02 pu. The speed appears to reach steady state after a number of oscillations. The decreases in the rotor speed are observed before going back to its state when the load is increased. An appropriate setting of the governor parameters (PID) would make the speed close to the synchronous speed. The simulation outcomes are

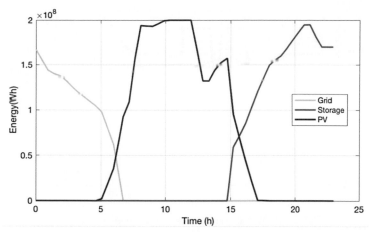

FIG. 5.11 Energy produced to match the load demand.

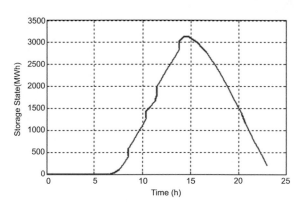

FIG. 5.12 Energy state of the gravity energy storage.

comparable with the results obtained by previous work presented in literature [7,22,40,59].

HYDRAULIC MODELING OF GRAVITY STORAGE

Dynamic Hydraulic Model

This section presents the mathematical models of system different parts, including the system hydraulic and dynamic behavior. The first step is to analyze the equilibrium forces of the piston-container assembly when it is in its static state. This is followed by a dynamic investigation of the system behavior. Pressure drop and friction effect are taken into consideration in the proposed model. Fig. 5.19 shows the system parameters as well as the applied forces during the discharging phase of gravity ESS. These parameters are used to derivate the model governing equations.

There is a significant interference between the different parts of the system. The variation of the valve slider position influences the motion of the piston inside the container, which also depends on the liquid compressibility, the flow inside the chambers, the flow rate, and the difference in pressure. The valve is responsible for the control and regulation of all the mentioned parameters. If a single parameter changes, all the other parameters will be influenced as well. Fig. 5.20 shows the link between the different parts of the hydraulic system. This model is highly coupled as a result of the dependency of the parameters on each other.

Volume dynamics

The water volume initial situation in the static mode $(x_p = 0)$ are

$$\begin{cases} V_{1,0} = 0 \\ V_{2,0} = (H_C - H_p)A_2 \end{cases} \quad (5.31)$$

The volume dynamics of the reservoirs is given by

$$\begin{cases} V_1(x_p) = V_{1,0} + x_p\dfrac{\pi d^2}{4} \\ V_2(x_p) = (H_C - H_p)A_2 - x_p\dfrac{\pi d^2}{4} \end{cases} \quad (5.32)$$

Pressure dynamics

Mass conservation equations are used to demonstrate the container's behavior. The mass of substances does not change in the hydraulic system (Eq. 5.33).

$$\dot{V} = \sum Q \quad (5.33)$$

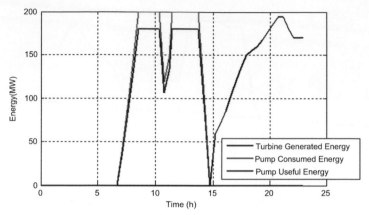

FIG. 5.13 Storage charging and discharging modes.

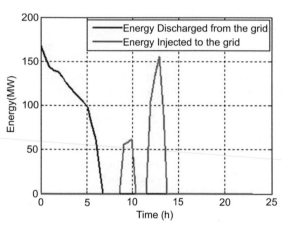

FIG. 5.14 Grid-exchanged energy.

The mass conservation principle is expressed as:

$$\dot{m} = \rho\dot{V} = \rho vA = const \qquad (5.34)$$

The conservation of fluid mass is given by:

$$\begin{cases} \dfrac{dm_1}{dt} = \rho_W(Q_1 - Q_L) \\[2mm] \dfrac{dm_2}{dt} = -\rho_W(Q_2 + Q_L) \end{cases} \qquad (5.35)$$

The fluid bulk modulus is taken into account in this study as the system studied operates in high-pressure conditions. Ignoring the aforementioned effect, it could compromise the system response behavior. A hydraulic system operating in high pressure requires the use of large amount of energy to compress the liquid. As a result, the response of the system is

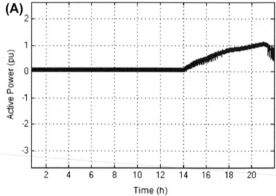

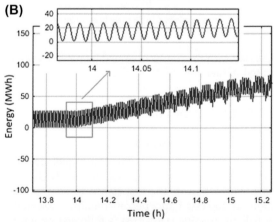

FIG. 5.15 **(A)** Active power of the generator in pu and **(B)** synchronous machine produced energy, Wh.

delayed. Piston motion cannot begin until the fluid has been properly compressed. Moreover, the energy stored may cause the piston to move even if the valve is closed [66].

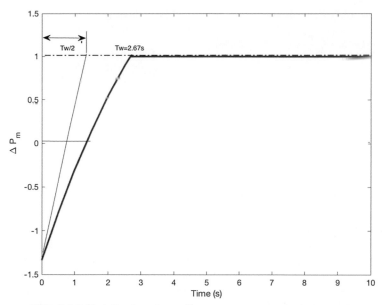

FIG. 5.16 Variation in gate position versus power output response.

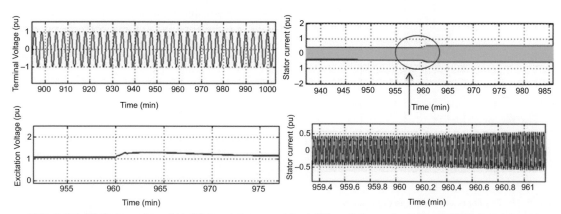

FIG. 5.17 **(A)** Stator voltage (V_a), **(B)** the stator current (I_{abc}), **(C)** excitation voltage (V_f), and **(D)** closer view of the stator current.

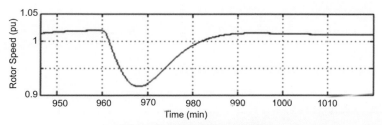

FIG. 5.18 Rotor speed variation.

The fluid density is formulated as follows:

$$\rho_W(p) = E_W(p)\frac{\delta\rho_W(p)}{\delta p} \tag{5.36}$$

A number of researches have been able to estimate bulk modulus of hydraulic fluids such as Jelali el al. and Totten et al. [66,67]. Experimental determination of several parameters is required for most of proposed equations. In this study, literature review enabled the estimation of the fluid bulk modulus. To express the bulk modulus of liquid elasticity, Eq. (5.37) is used:

$$E_W(p) = \frac{1}{2}E_0\log_{10}\left(k_1\frac{P}{P_{odn}} + k_2\right) \tag{5.37}$$

where P is pressure; E_0 is the nominal water compressibility modulus; E_W is the water compressibility modulus; and P_{odn} is reference pressure.

The following formula is obtained after combining Eqs. (5.35) and (5.36):

$$\begin{cases} \dfrac{dP_1}{dt}\dfrac{V_1\rho_W}{E_W(p)} + \dfrac{dV_1}{dt}\rho_W = \rho_W(Q_1 - Q_L) \\[2mm] \dfrac{dP_2}{dt}\dfrac{V_2\rho_W}{E_W(p)} + \dfrac{dV_2}{dt}\rho_W = -\rho_W(Q_2 + Q_L) \end{cases} \tag{5.38}$$

The expression of the derivative of pressure is

$$\begin{cases} \dfrac{dP_1}{dt} = \dfrac{E_W(p)}{V_1(x_p)}\left(Q_1 - Q_L - A_1\dfrac{dx_P}{dt}\right) \\[2mm] \dfrac{dP_2}{dt} = -\dfrac{E_W(p)}{V_2(x_p)}\left(Q_2 + Q_L - A_2\dfrac{dx_P}{dt}\right) \end{cases} \tag{5.39}$$

Piston motion dynamics

The motion of the piston is related to the pressure applied on its edges. The use of Newton's law in the equation of the piston motion gives

$$m\ddot{x}_P = F_1 - F_2 + F_g - F_f \tag{5.40}$$

where m is the piston mass; $\ddot{x}_P$ is the piston acceleration; F_1 and F_2 are the applied pressure force applied in chamber 1 and chamber 2, respectively; F_g is the gravitational force; and F_f the and friction force.

The expressions of piston motion including its acceleration, velocity, and position are

$$\begin{cases} \ddot{x}_P = \dfrac{1}{m}[(P_1A_1 - P_2A_2 + mg - F_f] \\[2mm] \dot{x}_P = \dfrac{\Delta x_P}{\Delta t} \\[2mm] x_P = \displaystyle\int \dot{x}_P \end{cases} \tag{5.41}$$

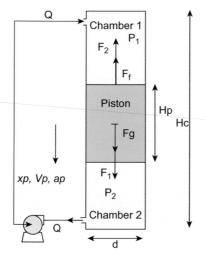

FIG. 5.19 Gravity energy storage parameters.

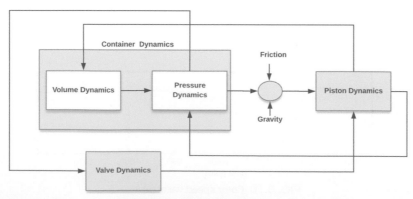

FIG. 5.20 Flowchart of the model.

Friction is considered a crucial aspect of piston motion [68]. The contact between the sealing, the container sides, the water flow, and the return conduit causes friction forces, which influence the dynamic behavior of GES. Usually, friction in hydraulics relies on several variables such as surface, materials, roughness, pressure difference, tolerances, and temperature; in addition to the seal characteristics such as its size, material, shape, and wear. Hydraulic systems are ordinarily described as having high dry friction forces and complex behavior because of the tight sealing they use [69]. The friction aspect is modeled in this work as an external disturbance. To calculate this force, the friction factor is multiplied by the force of the seal acting on the container's wall. Because it is hard to determine the value of the friction factor without experimental measurements, an estimated value of 0.1 is used [70]. This latter has been identified experimentally by the use of reciprocating seals in a container made of steel.

$$F_S = \mu F_N \tag{5.42}$$

To formulate the force acting on the container sealing by the initial deformation of the O-ring, Lindley's expression is used [71]. The normal force applied to the cylinder is given by:

$$F_N = b\pi D(P'_1 + \Delta P_f) \tag{5.43}$$

where b represents the contact width, D is the O-ring mean diameter, (ΔP_f) corresponds to the pressure difference in the system, and P'_1 is the contact pressure due to the initial compression. This latter is expressed as:

$$P'_i = d_S E(1.25C^{1.25} + 50C^6) \tag{5.44}$$

The fractional compression C is equal to the fraction of diametric compression (a). It is assumed, based on literature, that the initial fractional compression is equal to 25%. A less friction and stricter tolerances are represented by lower fractional compression. E represents the O-ring material elastic modulus. d_S is the O-ring cross-section diameter.

Yokoyama et al. derived an empiric equation (Eq. 5.45) for determining the contact pressure caused by the system pressure difference (ΔP_f) [72]:

$$\Delta P_f = \alpha \frac{P'_1(d\pi/2 - a)}{a} \tag{5.45}$$

where α is the conversion coefficient; it is assumed equal to 0.4 [73].

Eq. (5.46) as well as the Hertz theory is used to determine the contact width h [73]:

$$b = \sqrt{\frac{6P'_1 d_S}{\pi E}} \tag{5.46}$$

The presented equations do not take into account the time aspects such as relaxation. This study assumes that the effective contact stress does not change. Because of the low sensitivity of the sealing used, temperature effects are also neglected.

Hydraulic losses

Energy losses significantly affect the overall energy efficiency of the storage. In the studied gravity ESS, hydraulic loss should be determined as there is a loss of energy within the moving fluid in the container and the return pipe. These losses are due to friction within the moving fluid or in the pipe walls. In addition, fluid viscosity has also an impact on the system dynamics and efficiency. It may likewise result because of physical elements used by the system such as the valve or the pipe fittings. The aforementioned losses are classified as major and minor losses.

$$H_L = h_{Lmajor} + h_{Lminor} \tag{5.47}$$

where h_{Lmajor}, represents the system major losses that are caused by friction. Darcy-Weisbach equation (Eq. 5.48) is used to determine this type of hydraulic losses. This is used for incompressible and steady flow.

$$h_{Lmajor} = f * \frac{L}{D'} * \frac{V^2}{2g} \tag{5.48}$$

where f is the friction factor, L is the pipe length (m), D' is the pipe inside diameter (m), V is the average water velocity (m/s), and g is the gravitational acceleration (m/s^2).

For turbulent flow, Eq. (5.49) is used to calculate the friction factor [74].

$$f = \frac{1.325}{\left[\ln\left(\frac{e}{3.7D'} + \frac{5.74}{Re^{0.9}}\right)\right]^2} \tag{5.49}$$

where e is the relative roughness and is equal to $e = \frac{\varepsilon}{D'}$; v is the water kinematic viscosity, and Re is the Reynolds number $Re = \frac{VD}{v}$.

h_{Lminor} is the minor losses caused by the pipe bends, fittings, valves, and others. Each of these has a different loss coefficient (K). These losses are relative to the square of flow velocity. Eq. (5.50) is used to determine the hydraulic minor losses of the system [75]:

$$h_{Lminor} = K_L \frac{V^2}{2g} \qquad (5.50)$$

K_L is the loss coefficient. Eq. (5.51) is used to convert the water head (H_t) to pressure head (P_H).

$$H_t = P_H \frac{10}{SG} \qquad (5.51)$$

Valve dynamics

The opening of the valve initiates the discharge process of the system. The valve is responsible for controlling the fluid flow within the container's chambers. This flow is directed by its opening and closing. Therefore, the governing equation used to model the relieve valve depends on the pressure drop and the flow rate; this is expressed as:

$$P_2 - P_1 = R_v Q_2{}^2 + I \frac{dQ_2}{dt} \qquad (5.52)$$

The pressure in both chambers is represented by P_A and P_B. This latter is dependent on the flow resistance of the valve R_v and the pipe inductance (I), which is expressed as:

$$I = \frac{\rho L_T}{A_T} \qquad (5.53)$$

A_T the cross-sectional area of the return pipe and L_T is the pipe length.

The valve flow resistance is expressed as:

$$R_v = \frac{1}{2} \rho C_d \frac{1}{A_O{}^2} \qquad (5.54)$$

C_d represents the discharge coefficient of the valve and A_O is the valve opening area.

Simulink was used to solve the presented equations of the different hydraulic components. One of the aims of this proposed model is to visualize and analyze the hydraulic behavior of GES. The system parameters that can be identified as a function of time using the proposed model include chamber pressure, flow rate, volume, as well as piston position motion (acceleration, velocity, and position). To guarantee an accurate operation of the system, a control and management system is added to the model.

Control and management system

The role of the control and management system is to properly monitor the piston position and the velocity inside the container. The double integrator and the saturation block used by Simulink do not correctly represent the physical boundaries of the container. In other words, reaching the upper saturation limit of the piston velocity does not mean that the integration block will stop operating. That is why, it is important to model a control system. This latter allocates a zero to the velocity when the boundary limitations of the container are reached by the piston. The control system model is shown in Fig. 5.21.

Simulink Models

System-level model

The system level-model incorporates the different described governing equations. This high-level model of the system is shown in Fig. 5.22. The different submodels are linked to each other to develop the system-level model. The output of the control valve is the input of the piston and the container blocks. This output is used by the valve as input. The piston dynamics block output the position, velocity, and acceleration, which are used by the container block as input. The piston block inputs include both the pressure and the flow rate.

Valve model

The described equations were used to develop the sub-blocks of the system-level model. The valve dynamics' block diagram is shown in Fig. 5.23. In this model, the pressure in the container is considered as input, whereas the flow rate represents its output. For modeling the valve dynamics, it is necessary to make use of the return pipe geometry and the valve flow characteristics.

Pump model

In case the system makes use of separate pump-turbine, the most appropriate type of pumps used in this specific system configuration is centrifugal pump. This latter has been selected based on the system characteristics including the flow rate, the pressure head, and the power. SimHydraulics blocks have been used to develop the pump model as shown in Fig. 5.24. The aforementioned Simulink block package is specifically used to model hydraulic components. This includes sensors, electromotor subsystem, and a centrifugal pump. The control signal and the rotational speed are used by the pump as input, whereas the flow rate is

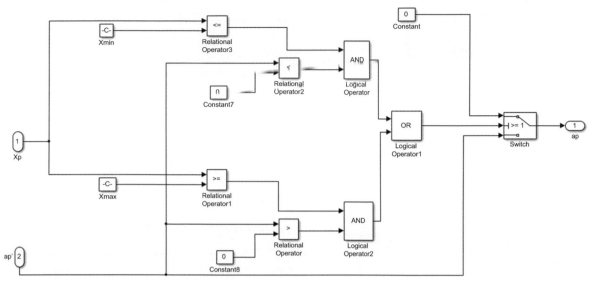

FIG. 5.21 Control and management system.

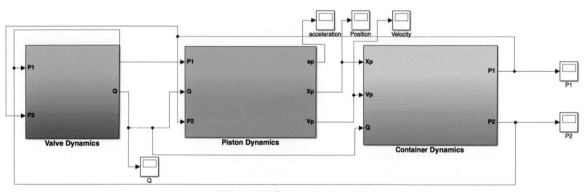

FIG. 5.22 System-level model.

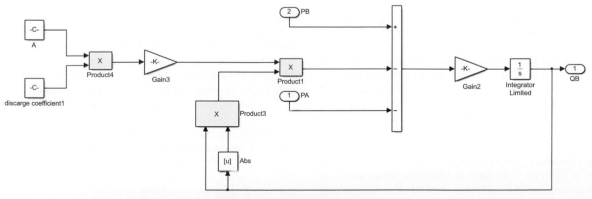

FIG. 5.23 Valve simulink model.

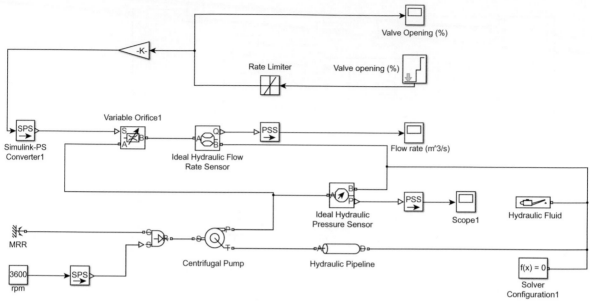

FIG. 5.24 Pump simulink model.

provided to the system as output. The subsystem of the electroengine compromises an ideal angular velocity source; linked to a mechanical and velocity reference blocks. The rotational speed is provided to the pump shaft by the subsystem. In addition, the model also incorporates hydraulic fluid reference, variable orifice, and a solver configuration. For measuring the system pressure and flow rate, sensors are incorporated to the model. The conversion of SI units to hydraulic units is done by the use of grains, as illustrated in Fig. 5.24.

Case Study Analysis

An interconnection of the different submodels, which have been described above as differential equations, gives a complete model. It is crucial to determine the different parameters of the model to perform a simulation study. These can be obtained from the specification of each component or through theoretic calculations and experimental measurement. Tables 5.2 and 5.3 show the system dimensions used in this case study. Eqs. 6.42–6.46 were used to estimate the frictional force in the system, whereas the liquid bulk modulus data were obtained from literature.

Simulation Results

The obtained piston motion results are shown in Fig. 5.25–5.26. The time it takes the piston to reach the bottom of the container is 368 s. It is shown that the velocity of the piston is constant throughout its

motion. The simulation stops as soon as the piston reaches the upper limit of the container.

Fig 5.27 compares the water volume in chambers 1 and 2 with respect to time as well as the system flow rate. It is shown that when the volume in chamber 1 increases, the water volume in chamber 2 decreases in parallel. In this case, the time it takes chamber 2 to discharge its volume (chamber 1 to fill up) is equal to 368. The measured flow rate through the valve is shown in Fig. 5.28 and is equal to $Q = 3.4 \times 10^{-4}$ m^3/s.

Fig. 5.29 illustrates the pressure variation of both chambers. It is noticed that while chamber 1 is empty chamber 2 is full of water because the piston is placed in the upper side of the storage container. Consequently, as it is shown in the graph, the pressure in chamber 2 is equal to 43.18 kPa ($P_2 = 43.18$ kPa) while it is negligible in chamber 1. The pressure in chamber 2 is caused by the piston and the water column forces. These latter are less than the piston pressure force. As a remark, during the discharging process of the storage system, the pressure of chamber 2 remains constant.

A closer view of the pressure graph is shown in Fig. 5.30 to analyze the system response due to the opening of the valve. The dynamic behavior of the system is dependent on this latter as they are in charge of converting the control signal into water flow. Typically, valves are characterized by a fast response. During this simulation, the complete opening of the valve is achieved quickly in about 0.82 s. A change in the

TABLE 5.2
Parameters of Gravity Energy Storage Components.

Component	Container		Piston			Return Pipe	
Parameter	Height	Diameter	Height	Diameter	Mass	Length	Cross-Sectional Area
Value	2 m	0.4 m	1 m	0.4 m	988 Kg	2 m	0.0005 m^2

TABLE 5.3
Sealing and Valve Components.

Component	Valve			Sealing		
Parameter	Opening Area	Discharge Coefficient	O-Ring Cross-Section Diameter	O-Ring Mean Diameter	E Elastic Modulus of the O-Ring Material	Fractional Compression
Value	0.00006 m^2	0.611	8 mm	400 mm	5.52 MPA	25%

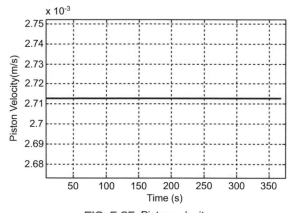

FIG. 5.25 Piston velocity.

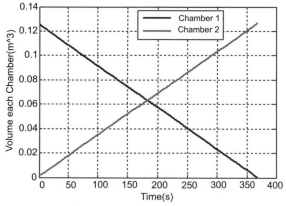

FIG. 5.27 Volume within chamber 1 and 2 during the system discharging phase.

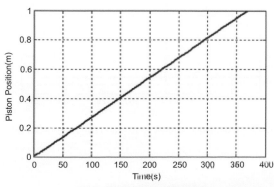

FIG. 5.26 Piston position.

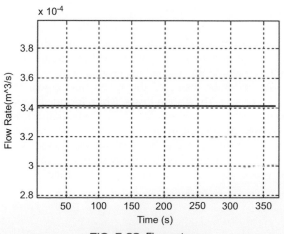

FIG. 5.28 Flow rate.

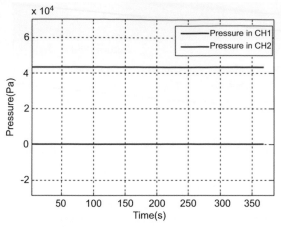

FIG. 5.29 Pressure within chamber 1 and 2.

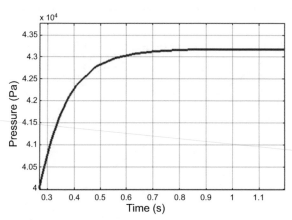

FIG. 5.30 Chamber 2 pressure variation with a closer view.

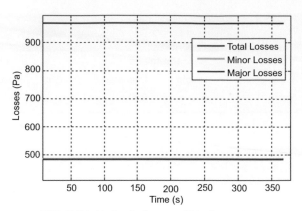

FIG. 5.31 Hydraulic losses within the system.

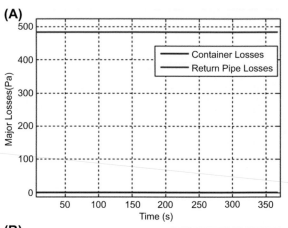

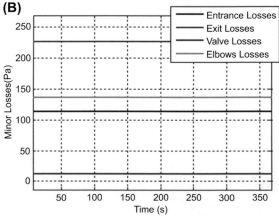

FIG. 5.32 **(A)** Major hydraulic losses and **(B)** minor hydraulic losses.

pressure is observed as the valve is being opened. During this process, a rapid change is noticed before reaching a stable value of 43 kPa. This pressure is achieved in less than 1 s as the valve is opened fully. A constant pressure is obtained because it is a function of the piston velocity and the water flow rate, which remain constant throughout the discharging process of the system.

The hydraulic losses of the system are shown in Fig. 5.31−5.32. The system losses, which include minor and major losses, are about 968 Pa. Major hydraulic losses caused by friction in both the return pipe and the container are estimated to be equal 482 Pa. These latter types of losses depend on the geometry of the pipe and the water flow velocity. Because of the high water velocity in the return pipe, frictional losses in this pipe are significantly higher than those in the container. They are estimated to be equal to 482 Pa in the return pipe, while they are negligible in the container (0.0006 Pa). Minor losses, on the other hand, are associated with flow losses through the valve, pipe elbows, exit, and entrance. These are estimated to be equal to 486 Pa. The highest minor losses occur in the pipe exit, followed by losses in pipe elbows and entrance, respectively. Losses in the valve are considered as the smallest minor losses. This type of losses depends on the component physical characteristics and geometry, which are represented by a loss coefficient k.

Fig. 5.33 shows a change in the pump flow rate as a function of time due to the gradual opening of valve. As this latter is being opened, the flow rate increases step by step. Once the valve is completely opened, the flow rate is stabilized in less than 1 s.

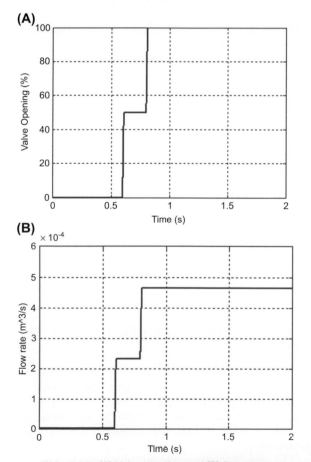

FIG. 5.33 (A) Valve opening and (B) flow rate.

EXPERIMENTAL TESTS
Experimental Validation

The results obtained in the performed simulation are compared with experimental tests to verify the effectiveness of the model. A prototype named "Powertower 1" was performed by Aufleger et al. at the University of Innsbruck [76]. A second case study has been carried out because of the difference in the system dimensions between the experimental test and the first performed case study.

The available data associated with the geometry of the experimental test are limited. These include the container height and diameter, which are equal to 2.2 and 0.6 m, respectively. The material used to construct the piston, which has a mass of 1500 kg, is stainless steel piston. The different dimensions of the prototype are shown in Table 5.4. The presented data are used to derive some of the missing input data using physics' formulas (see Table 5.5).

Estimations of those parameters had to be done to carry out the simulation. The characteristics of the valve used have been estimated based on literature review such as the discharge coefficient and the opening area. Estimations of some parameters associated with the O-ring,and the sealing have been used to determine the friction losses. The diameter of the sealing cross section is 8 mm while the mean diameter is 600 m. The O-ring material used has an estimated elastic modulus of 5.52 MPA.

The simulation-obtained results for the system discharging process are presented in Table 5.6. The time it takes the piston to reach its maximum position is 336 s. It travels with a velocity of 4.52×10^{-3} m/s. The system pressure caused by the piston and the water column is about 40.6 kPa. Minor and major losses are also estimated in this case study.

The results obtained from the simulation and the experimental tests are compared with validate the proposed model. Energy storage charge and discharge cycles, used as a function of time, are shown in Fig. 5.34. The small disparity between the stimulated and the experimental models can be explained by the estimations of the system geometry. The available data about the system storage rated power and capacity were used to calculate the prototype discharging time and has been found equal to 315 s. The simulated model outcomes show that the time necessary for the piston to travel to its final position is 336 s during a half-cycle. The disparity between the simulated and the experimental results was expected because system losses were estimated using different formulas. It should be noted that an approximation of the valve

TABLE 5.4
Experimental Data.

	Container's Height	Container's Diameter	Piston Mass	Density	Pressure	Storage Capacity	Storage Power
Experimental model	2.2 m	0.6 m	1500 kg	7850 kg/m^3	4 mWs	3.5 Wh	40 W

TABLE 5.5
Model-Derived Data.

	Piston Height	Piston Diameter	Water Height	Return Pipe Height
Value	0.67 m	0.6 m	1.524 m	2.2 m

TABLE 5.6
Simulation-Obtained Results.

	Piston Velocity (m/s)	Discharge Time (s)	Pressure (kPa)	Flow Rate (m^3/s)	Friction Loss (kN)	Minor Losses (Pa)	Major Losses (Pa)
Value	4.52×10^{-3}	336	40.6	1.27×10^{-3}	6.584	6838	5036

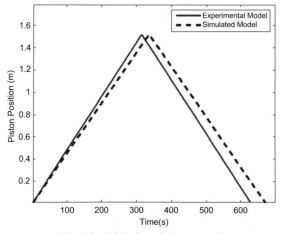

FIG. 5.34 Variation of piston position.

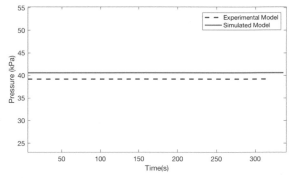

FIG. 5.35 Pressure head of the experimental and simulated models.

area was used. The obtained discharge time could be lower if a larger opening area of the valve had been used. Consequently, it is necessary to perform a sensitivity analysis to analyze the impact of varying the valve size on the simulation results.

The pressure head of the experimental and simulated models are 39.2 kPa and 40.6 kPa, respectively. Fig. 5.35 shows a comparison of the pressure in both models. A small difference is obtained between the two models.

The area that is above the piston is named chamber 1, while chamber 2 represents the area that is below the

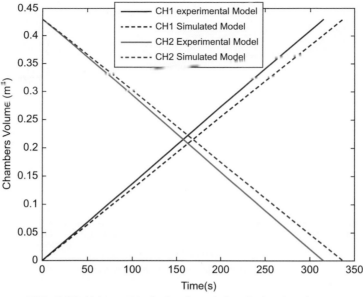

FIG. 5.36 Volume of both chambers during discharging phase.

TABLE 5.7
Results Comparison.

	Time	Pressure
Experimental model	315	39.2
Simulated model	336	40.6
% error	6.2%	3.4%

piston. Fig. 5.36 shows the charging and discharging of the systems' chambers. When the piston is moving downward, the volume of chamber 1 increases. At the same time, the volume of chamber 2 decreases. When chamber 2 is completely empty, the piston has reached its maximum position. This occurs at 315 s for the experimental model, while it occurs at 336 s for the simulated one. The experimental model takes less time to store and discharge energy than the simulated one.

The simulated and the experimental results are close to each other with a small percent errors. Table 5.7 shows the % errors of the compared results. These errors may result because of the estimated input parameter, especially those related to the valve size. Therefore, it is necessary to carry out simulations with relatively small and large variations of the valve parameters to analyze the influence of the valve dimensions on the

storage discharge time and pressure. The impact of this variation on the system performance is shown in Fig. 5.37.

The results demonstrate that the valve size has a significant effect on the system discharge time and pressure. An increase of the pressure is shown along with a decrease in the discharge time as the valve area is increased. A smaller valve size will lead to a decrease in the % error of the pressure between the simulated and the experimental models. Conversely, this change will results in a rise of the % error of the discharge time.

The sensitivity analysis shows that the estimations used for the valve size influence the dynamic behavior of the system. The results of the simulated model and the experimental one are relatively close when the opening area of the valve, which is equivalent to 2.32×10^{-4} m^2 (see Table 5.8).

Sensitivity analysis

A sensitivity analysis is carried out in this section to examine the impact of the different assumptions made on the model output. These assumptions will be explored further to investigate their influence on the behavior of the system. Some of the estimated parameters include the valve discharge coefficient, bulk modulus, friction, piston mass, along with hydraulic loss coefficients. This analysis has an objective of testing the strength of the simulation results and

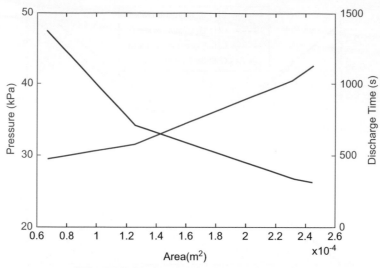

FIG. 5.37 Discharge time and pressure impact.

TABLE 5.8 % Errors With the Variation of Valve Size.			
Valve Diameter (in)	**Valve Opening Area (10^4 m^2)**	**Pressure % Error (%)**	**Discharge Time % Error (%)**
0.37	0.67	34	77
0.5	1.26	32	55
0.68	2.32	3.4	6.2
0.7	2.45	7	1.2

reducing the uncertainty of the model. This is done through the identification of the different parameters that have a direct effect on the simulation outputs.

The system hydraulic behavior is one of the main priorities of the proposed model. To investigate the system dynamics, it is very important to perform a sensitivity analysis on the discharge coefficient of the valve and the system losses.

The mass of the piston can be different based on the specific design of the system. Therefore, a sensitivity analysis is carried out on the piston mass to visualize the influence of the variation of this parameter on the system behavior. This study could also result in an optimal design of the system.

Several authors have proposed different methodologies to model the fluid empiric bulk modulus. The use of a specific method may have an impact on the simulation results. Therefore, it is necessary to examine the effect of the selected bulk modulus on the system

behavior. Furthermore, friction force is another parameter that is affected by several undefined conditions. Therefore, examining if this latter would affect the simulation outputs is very crucial. In this sensitivity analysis, different scenarios have been studied with different parameters. An increase and a decrease of the aforementioned parameters by 20% and 40% is used in this analysis.

The simulation results are presented in Table 5.9. A proportional change is obtained in the flow rate as the different parameters, which include the bulk modulus, the discharge coefficient of the valve, and hydraulic losses, are varied. For instance, the flow rate changes by approximately 20% as a result of a 20% variation in the aforementioned parameters. As for the pressure, a very low variation is observed even for a high change of these parameters; this variation is less than 2%. Conversely, the impact of the piston mass on the system pressure is significant, while it has a low effect on the

TABLE 5.9
Impact of Pressure and Flow Rate.

	% CHANGE IN FLOW RATE				% CHANGE IN PRESSURE			
	Discharge Coefficient	Hydraulic Losses	Piston Mass	Fluid Bulk Modulus	Discharge Coefficient	Hydraulic Losses	Piston Mass	Fluid Bulk Modulus
Actual Value	3.40×10^{-4} m³/s				4.31 kPa			
Increase of 20%	20%	21.1%	17%	20%	0.92%	1.6%	36%	2%
Increase of 40%	41.4%	42.9%	31.7%	41.7%	2%	3.94%	72.85%	2%
Decrease of 20%	−20.2%	−20.2%	−20.2%	−20.2%	−0.69%	−0.9%	−36.42%	−0.69%
Decrease of 40%	−40%	−40%	−48.2%	−40%	−1.39%	−1.62%	−73%	−1.39%

flow rate. The pressure level is increased by the use of piston with larger mass. Therefore, a change in the mass of the piston results in a significant variation in the system dynamics.

The flow hydraulic losses have the most significant impact on the system flow rate. This is followed by the fluid bulk modulus. The mass of the piston and the valve discharge coefficient have less impact on this flow rate. The chamber pressure is, on the other hand, altered by a change in the mass of the piston. The impact of the hydrolosses, the discharge coefficient of the valve, and the effective bulk modulus is less significant.

This analysis has shown that system losses are important parameters that should be considered carefully for the determination of the flow rate. Conversely, the piston mass is also considered as an essential parameter as it defines the pressure of the system. Although some of the investigated aspects do not actually have significant impact on the storage behavior, a sensitivity analysis should be carried out to examine their impacts. In addition, valid estimations of diverse parameters are crucial to prevent errors associated with the performance of the system dynamic.

As explained before, it is hard to approximate the friction aspects without running experimental tests. However, it was possible to estimate this force with the use of formulas available in literature. It is known that piston velocity is influenced by the friction force. In addition, it is shown in Table 5.10 that when this latter force is increased by 20%–40%, its effects on

TABLE 5.10
Impact of Friction Force Variation.

(% Change)	Piston Velocity	Flow Rate	Pressure
Actual value	2.71	3.40	4.31
Increase of 20%	9.96%	10.2%	−15.5%
Increase of 40%	15.8%	15.8%	−31.5%
Decrease of 20%	−14%	−13.8%	15.5%
Decrease of 40%	−31.3%	−31.1%	31.5%

the piston velocity decrease. A similar impact is obtained for the flow rate. In contrast, friction has a major impact on the system pressure. This latter is more significant than the impact provoked by the other studied parameters, expect for the mass of piston. Therefore, the system dynamic behavior is influenced considerably by the approximations made to estimate the friction force.

The estimations used to approximate some of the input parameters may influence the system dynamics as explained before. Therefore, a careful approximation of these latter is necessary to build a robust model. By the proper use of formulas and assumptions available in literature, it is possible to appropriately estimate the different investigated parameters, which include, in this case, the valve parameters, the hydraulic losses, the friction force, and bulk modulus.

Design Considerations

The fluid in chamber 2 is pressed by the piston; this will cause the water to flow toward chamber 1. Because the fluid always chooses the path that has the least resistance, a sealing system between the container and the piston is added with the purpose of preventing the flow of water between chamber 1 and 2. In addition, the sealing is responsible for the guidance of the piston movement. A watertight closure is used to prevent the fluid that is under pressure from escaping between the chambers. The sealing material type that is commonly used is plastic polymer; this contains rubber O-ring and composite seals, which incorporate a steel spring and a plastic polymer. The use of self-energizing O-rings makes these types of sealing watertight even under high-pressure conditions. To accommodate the ring deformation, this should be made from materials that have a high Poisson ratio and are incompressible. It is important to have a frequent replacement of the seals used to prevent its wear.

The downward and upward movement of the piston causes friction between the piston sealing and the container's walls. The total efficiency of the storage system is influenced considerably by this friction. This friction might be minimized by the use of sealing made of plastic polymers. Polytetrafluoroethylene (PTFE) is an example of plastic polymers that might be used because of its low elasticity and high wear resistance. To deal with the low elasticity of a sealing material, it is necessary to make use of an energizing O-ring or an integrated steel spring. To decrease the seal wear and friction between the piston sealing and the container's wall, this latter should be characterized by a low surface roughness. In this situation, the surface roughness should be lower than 0.25 μm to ensure a proper functioning of the PTFE sealing [77]. Generally, the friction force depends on the container surface roughness, the piston velocity, the pressure within the system, and the seal characteristics, which include its size, material, shape, and wear.

The system round-trip efficiency is affected by the friction losses. This latter can be expressed for a storage charge/discharge cycles as:

$$E_f = 2\mu \int_{Z_2}^{Z_1} F_N dz \qquad (5.55)$$

The pipe losses at the inlet and outlet, as well as friction losses are dependent on the velocity of the flow as given in Eqs. (5.48)–(5.50). Larger pipes should be used to reduce these losses. In addition, there is a loss of energy, as much as the pipe is restrictive. Reducing the pipe length and number of elbows is also necessary to minimize the hydraulic losses in the system. In addition, materials with low surface roughness should be used. The losses can also be reduced by minimizing the friction coefficient of the pipe and the valve. Extra power should be provided by the pump to overcome these hydraulic losses. Therefore, it is essential to optimize the dimensions of the piping system and to properly choose the pump equipment, which would fulfill the condition of the system to increase the storage operational capability.

The pressure head of GES is considerably influenced by the mass of the piston; therefore, it impacts also the storage capacity. As presented in Eq. (5.56), the capability of GES is a function of the piston density. Designing a piston with a height equivalent to half of the container's height provides an optimal configuration of the piston size.

$$E_c = \frac{1}{4}\pi(\rho_p - \rho_w)d^2 H_P gz\eta \qquad (5.56)$$

CONCLUSION

This chapter has examined the behavior and performance of GES coupled to a renewable energy plant. A dynamic model has been developed for the aforementioned system with the use of an electric grid. This latter has been used to dispatch energy between the aforementioned system and a residential load. GES system was responsible for balancing and meeting the load energy demand by discharging energy and storing the PV excess power. Matlab/Simulink has been used to develop the different models of the system components.

To examine the performance of the hybrid system, a case study was carried out to validate the effectiveness of the proposed models. The obtained outcomes show that the model is capable of responding to load demand in accordance with the presented management strategy. The dynamic modeling of the mechanical equipment of GES enabled for the visualization of the system behavior and its response to energy demand. A load change has an impact on the generator parameters as indicated by the performed study. As soon as the load was varied, some oscillations were detected. These oscillations did take place in the excitation voltage, the generator's stator current, and the generated power. As for the terminal voltage, no oscillations were observed. However, when the system gets close to its stable state, low oscillations occurred. The system oscillations may be reduced by the appropriate tuning of the PID controller parameters. Similar mechanical equipment are used by pumped hydrostorage system and hydropower plants. The generator and the hydroturbine real performance and behavior are demonstrated by the developed model.

The hydraulic components of the storage system were modeled in this study. This model was also created using Matlab/Simulink by taking into consideration the system characteristics as well as the different properties of the fluid flow. With an aim of examining the behavior of this system, a case study was performed. The proposed model works as a virtual system representing GFS. The system behavior as a function of time was determined without experimentally testing the system. The storage characteristics such the piston motion, the discharge time, along with the chamber's volume, pressure, and flow rate are identified by this model. The obtained results demonstrate that the system discharge time is about 368 s. In addition, the pressure is equal to 43.1 kPa, whereas the flow rate has been found equal to 3.4×10^{-4} m^3/s. To validate the presented model, a comparison of the simulation results with experimental tests performed by other researchers has been done. The outcomes of this comparison have shown that the percent errors are low for both the pressure and the discharge time as they are equal to 3.4% and 6.2%, respectively. The obtained % error may be due to the estimated or the derived system parameters or because of the assumptions made in the simulated case study.

To assess the system response to a change in the parameters of the performed simulation, a sensitivity analysis was carried out in this work. Because some input parameters were approximated, it is essential to perform this study. In addition, to examine the influence of some estimated parameters on the pressure and the flow rate of the system, different situations with a 20%—40% increase and decrease of these parameters were studied. The obtained outcomes demonstrate that a change in the bulk modulus, the hydraulic losses, and the discharge coefficient of the valve results in a proportional change in the system flow rate. The piston mass has a considerable influence on the system pressure. This study has shown that it is important to correctly estimate the values of the investigated parameters; mainly those associated with losses, which consist of friction and hydraulic losses.

A number of design concerns were presented in this chapter. The simulation outcomes demonstrate that the efficiency of the system is dependent on the materials of the sealing system used. This latter can be improved by reducing the system hydraulic losses. This can be done through the proper selection and design of the piping

system, the valve, and the pump. In addition, because the piston mass has a considerable influence on the system pressure, dimensioning this latter in an optimal manner would results in a greater storage capacity. The discussed design considerations could be used to obtain a more energy-efficient system with minimized hydraulic losses.

NOMENCLATURE

a	Diametric compression (mm)
A	Water conduit area (m^2)
A$_1$	Area of chamber 1 (m^2)
A$_2$	Area of chamber 2 (m^2)
A$_t$	Factor that account for the different in per units
A$_O$	Opening area of the valve (m)
A$_T$	Cross sectional area of the return pipe (m^2)
b	Contact width (mm)
C	Fractional compression
C$_1$	Effective compliance of chamber 1
C$_2$	Effective compliance of chamber 2
C$_d$	Valve discharge coefficient
d	Diameter of the container and piston (m)
D	Mean diameter of the O-ring (mm)
D$_c$	Damping coefficient
D$'$	Inside diameter of the pipe (m)
d$_S$	Cross-section diameter of the O-ring (mm)
e	Relative roughness
E	Elastic modulus of the O-ring material (N/m^2)
E$_c$	Storage capacity
E$_D$(t)	Energy discharged from the storage at t (W)
e$_f$	Regulator output
E$_f$	Friction losses
E$_S$(t)	Energy stored at time t (W)
E$_0$	Nominal fluid compressibility modulus (kPa)
E$_W$	Liquid compressibility modulus (kPa)
f	Friction factor
F$_1$	Pressure forces applied by chamber 1 (Pa)
F$_2$	Pressure forces applied by chamber 2 (Pa)
F$_f$	Friction force (N)
F$_l$	Hydraulic loss (N)
F$_g$	Gravitational force (N)
F$_N$	Force of the seal against the cylinder wall (N)
F$_{net}$	Net force on the water in the conduit (N)
F$_S$	Static friction (N)
g	Gravitational acceleration (m/s^2)
G	Gate opening (m)
h	Head across the turbine (m)
H	Head at turbine gate (m)

h_{base}	Base head (m)
H_C	Height of the container (m)
H_d	Discharge head (m)
H_L	Total hydraulic losses (Pa)
h_{Lmajor}	Major hydraulic losses (Pa)
h_{Lminor}	Minor hydraulic losses (Pa)
H_P	Height of the piston (m)
H_s	Static head (m)
H_t	Water head (Pa)
I	Channel inductance ($Kg.m^2.s^{-2}.A^{-2}$)
K	Constant
K_L	Loss coefficient
Ke	Feedback gain
L	Water conduit length (m)
$L(t)$	Energy demand (load) at time t (W)
L_T	Length of the return pipe (m)
m	Piston mass (Kg)
P	Pressure (Pa)
P_1	Pressure in chamber 1 (Pa)
P_2	Pressure in chamber 2 (Pa)
P_C	Power consumed by the pump in (W)
P_d	Pressure at discharge (Pa)
Pgrid (t)	Power transferred from the grid to the load (W)
P_H	Pressure head (Pa)
P'_1	Contact pressure due to the initial compression of the O-ring (Pa)
$P_L(t)$	Power demand at time t; PD(t) is the power discharged from the storage system (W)
P_m	Turbine mechanical power (W)
P_{odn}	Reference pressure (Pa)
P_p	Peak nominal power based on 1 kW/m^2 radiation at standard test condition (W)
P_S	Suction pressure (Pa)
$P_S(t)$	Power stored from the PV system (W)
P_U	Useful power (W)
$PV(t)$	Power produced by the PV system (W)
Q	Flow rate (m^3/s)
Q_1	Flowrates in chamber 1 (m^3/s)
Q_2	Flowrates in chamber 2 (m^3/s)
$Q_{base.}$	Base flow rate (m^3/s)
Q_v	Flow of water (m^3/s)
q_{nl}	No load flow
Re	Reynolds number
Rv	Flow resistance of the valve ((N/m^2)/(m^3/s))
R_T	Temporary drop
s	Second
SG	Specific gravity of water
Sr	Available solar radiation for a particular location (Wh/m^2)
$S(t)$	Storage level at time t (Wh)
$S(t-1)$	Storage remaining energy at time (t-1) (Wh)
t	Time (s)

Te	Time constant (s)
Tg	Pilot valve droop (s)
T_P	Pilot valve time constant (s)
Tr	Dashpot time constant (s)
T_R	Reset time (s)
Tw	Water time constant (s)
v	Kinematic viscosity of water (Pa.s)
V	Average water velocity (m/s)
V_1	Volume in chamber 1 (m^3)
V_2	Volume in chamber 2 (m^3)
$V_{1,0}$	Initial volume in chamber 1 (m^3)
$V_{2,0}$	Initial volume in chamber 2 (m^3)
V_{fd}	Exciter voltage (V)
x_p	Piston position (m)
$\dot{x}_P$	Piston velocity (m/s)
$\ddot{x}_P$	Piston acceleration (m^2/s)
W_p	Maximum power of each module at STC (W)
z	Elevation height (m)
α	Conversion coefficient
δ	Self-discharge rate of the system
ΔP_f	Contact pressure due to the system pressure difference (Pa)
η	System efficiency
μ	Friction factor
ξ	Inclination and orientation correction factor of the solar system (W)
ξ_o	Discharge coefficient
ρ_p	Density of piston (kg/m^3)
ρ_w	Density of water (kg/m^3)
ω	Rotor speed (rmp)

REFERENCES

[1] Albadi MH, El-Saadany EF. Overview of wind power intermittency impacts on power systems. Electr Power Syst Res 2010;80(6):627–32.

[2] Smith JC, Milligan MR, DeMeo EA, Parsons B. Utility wind integration and operating impact state of the art. IEEE Trans Power Syst 2007;22(3):900–8.

[3] Beaudin M, Zareipour H, Schellenberglabe A, Rosehart W. Energy storage for mitigating the variability of renewable electricity sources: an updated review. Energy Sustain Dev 2010;14(4):302–14.

[4] Moura PS, De Almeida AT. The role of demand-side management in the grid integration of wind power. Appl Energy 2010;87(8):2581–8.

[5] Ferreira HL, Garde R, Fulli G, Kling W, Lopes JP. Characterisation of electrical energy storage technologies. Energy 2013;53:288–98.

[6] Maclay JD, Brouwer J, Samuelsen GS. Dynamic modeling of hybrid energy storage systems coupled to photovoltaic generation in residential applications. J Power Sources 2007;163:916–25.

[7] Li C-H, Zhu X-J, Cao G-Y, Sui S, Hu M-R. Dynamic modeling and sizing optimization of stand-alone photovoltaic power systems using hybrid energy storage technology. Renew Energy 2009;34:815—26.

[8] Jayalakshmi NS, Gaonkar DN, Balan A, Patil P, Raza SA. Dynamic modeling and performance study of a stand-alone photovoltaic system with battery supplying dynamic load. Int J Renew Sustain Energy Res 2014;4(3).

[9] Nemati M, Tenbohlen S, Braun M, Imran M. Development of generic dynamic models for distributed generators in microgrids. http://www.unistuttgart.de/ieh/forschung/veroeffentlichungen/2013_Nemati_Development_of_Generic_Dynamic_Models.pdf.

[10] Sarasua AE, Molina MG, Mercado PE. Dynamic modelling of advanced battery energy storage system for grid-tied AC microgrid applications, Chapter 12 energy storage — technologies and applications. book published by Intech https://doi.org/10.5772/52219.

[11] Nayeripour M, Hoseintabar M. A comprehesnive dynamic modeling of grid connected hybrid renewable power generation and storage system. Int J Model Optim April 2011;1(1).

[12] Gustavo Molina M. In: Brito AV, editor. Dynamic modelling and control design of advanced energy storage fo power system applications, dynamic modelling. InTech; 2010, ISBN 978-953-7619-68-8. Available from: http://www.intechopen.com/books/dynamic-modelling/dynamic-modelling-and-control-designof-advanced-energy-storage-for-power-system-applications.

[13] Dugan R, Taylor J, Delille G, Storage simulations for distribution system analysis. In: Paper 1340, 22nd international conference on electricity distribution Stockholm, 10—13 June 2013.

[14] Bazan P, German R Hybrid simulation of renewable energy generation and storage grids. In: Laroque C, Himmelspach J, Pasupathy R, Rose O, and Uhrmacher AM, eds. Proceedings of the 2012 winter simulation conference .

[15] DiOrio N, Dobos A, Janzou S, Nelson A, Lundstrom B. Technoeconomic modeling of battery energy storage in SAM. September 2015. Technical Report NREL/TP-6A20-64641.

[16] Shaifi Asl S-M, Rowshanzamir S, Eikani M-H. Modeling and simulation of steady-state and dynamic behavior of a PEM fuel cell. Energy 2010;35:1633—46.

[17] Douglas T. Dynamic modelling and simulation of a solar-PV hybrid battery and hydrogen energy storage system. J Energy Storage 2016;7:104—14.

[18] Yigit T, Selamet O-F. Mathematical modeling and dynamic Simulink simulation of high-pressure PEM electrolyzer system. Int J Hydrog Energy 2016;41:13901—14.

[19] Harish V-S-K-V, Kumar A. A review on modeling and simulation of building energy systems. Renew Sustain Energy Rev 2016;56:1272—92.

[20] Wu J. Experimental investigation of the dynamic behavior of a large-scale refrigeration e PCM energy storage system. validation of a complete model. Energy 2016;116:32—42.

[21] Arabkoohsar A, Andresen G. Dynamic energy, exergy and market modeling of a high temperature heat and power storage system. Energy 2017;126:430—43.

[22] Zhu J, et al. Design, dynamic simulation and construction of a hybrid HTS SMES (high-temperature superconducting magnetic energy storage systems) for Chinese power grid. Energy 2013;51:184—92.

[23] European Commission. Energy storage-the role of electricity. 2017. Available at: https://ec.europa.eu/energy/sites/ener/files/documents/swd2017_61_document_travail_service_part1_v6.pdf.

[24] Maton J-P, Zhao L, Brouwer J. Dynamic modeling of compressed gas energy storage to complement renewable wind power intermittency. Int J Hydrog Energy 2013;38:7867—80.

[25] Mazloum Y, Sayah H, Nemer M. Dynamic modeling and simulation of an isobaric adiabatic compressed air energy storage (IA-CAES) system. J Energy Storage 2017;11:178—90.

[26] Saadat M, Shirazi FA, Li PY. Modeling and control of an open accumulator compressed air energy storage (CAES) system for wind turbines. Appl Energy 2015;137:603—16.

[27] De Jaeger E, Janssens N, Malfliet B, et al. Hydro turbine model for system dynamic studies. IEEE Trans Power Syst 1994;9:1709—15.

[28] Singh M, Chandra A. Modeling and control of isolated micro-hydro power plant with battery storage system. In: National power electronic conference, Roorkee, India; 2010.

[29] Guo A-W, Yang JD. Self-tuning PID control of hydro-turbine governor based on genetic neural networks. Adv Comput Intell 2007;4683:520—8.

[30] Pennacchi P, Chatterton S, Vania A. Modeling of the dynamic response of a Francis turbine. Mech Syst Sig Process 29, 107—19. https://doi.org/10.1016/j.ymssp.2011.05.012.

[31] Bhaskar MA. Non linear control of STATCOM. In: IEEE international conference on recent trends in information telecommunication and computing; 2010. p. 190—5.

[32] Sanathanan CK. Accurate low order model for hydraulic turbine-penstock. IEEE Trans Energy Convers 1987:196—200. EC-2.

[33] Hannett LN, Feltes JW, Fardanesh B. Field tests to validate hydro turbine-governor model structure and parameters. IEEE Trans Power Syst 1994;9:1744—51.

[34] Jiang CW, Ma YC, Wang CM. PID controller parameters optimization of hydro-turbine governing systems using deterministic-chaotic-mutation evolutionary programming (DCMEP). Energy Convers Manage 2006;47:1222—30.

[35] Liu YJ, Zhu XM, Fang YJ. Modeling of hydraulic turbine systems based on a Bayesian—Gaussian neural network driven by sliding window data. J Zhejiang Univ Sci C 2010;11:56—62.

[36] Chen D, Ding C, Ma X, Yuan P, Ba D. Nonlinear dynamical analysis of hydro-turbine governing system with a surge tank. Appl Math Model 2013;37:7611—23.

[37] Shen ZY. Hydraulic turbine regulation. Beijing: China Waterpower Press; 1998. in Chinese.

[38] Fang HQ, Chen L, Dlakavu N, et al. Basic modeling and simulation tool for analysis of hydraulic transients in hydroelectric power plants. IEEE Trans Energy Convers 2008;23:834–41.

[39] Inayat-Hussain JI. Nonlinear dynamics of a statically misaligned flexible rotor in active magnetic bearings. Commun Nonlinear Sci Numer Simul 2010;15:764–77.

[40] Nanaware RA, Sawant SR, Jadhav BT. Modeling of hydraulic turbine and governor for dynamic studies of HPP. IJCA 2012:0975–8887.

[41] Fang HQ, Chen L, Shen ZY. Application of an improved PSO algorithm to optimal tuning of PID gains for water turbine governor. Energy Convers Manage 2011;52: 1763–70.

[42] Das S, Pan I, Das S. Fractional order fuzzy control of nuclear reactor power with thermal-hydraulic effects in the presence of random network induced delay and sensor noise having long range dependence. Energy Convers Manage 2013;68:200–18.

[43] Zhu H, Huang WW, Huang GH. Planning of regional energy systems: an inexact mixed-integer fractional programming model. Appl Energy 2014;113:500–14.

[44] Zhu H, Huang GH. Dynamic stochastic fractional programming for sustainable management of electric power systems. Int J Electr Power Energy Syst 2013;53: 553–63.

[45] Ozdemir MT, Sonmez M, Akbal A. Development of FPGA based power flow monitoring system in a microgrid. Int J Hydrogen Energy 2014;39:8596–603.

[46] Chilipi RR, Singh B, Murthy SS, Madishetti S, Bhuvaneswari G. Design and implementation of dynamic electronic load controller for three-phase selfexcited induction generator in remote small-hydro power generation. IET Renew Power Generation 2014;8:269–80.

[47] Xu Y, Li ZH. Computational model for investigating the influence of unbalanced magnetic pull on the radial vibration of large hydro-turbine generators. J Vib Acoust Trans ASME 2012;134:051013.

[48] Kishor N, Singh SP, Raghuvanshi AS. Dynamic simulations of hydro turbine and its state estimation based LQ control. Energy Convers Manage 2006;47:3119–37.

[49] Zeng Y, Guo YK, Zhang LX, Xu TM, Dong HK. Nonlinear hydro turbine model having a surge tank. Math Comput Model Dyn Syst 2013;19:12–28.

[50] Zhang H, Chen D, Xu B, Wang F. Nonlinear modeling and dynamic analysis of hydro-turbine governing system in the process of load rejection transient. Energy Convers Manag 2015;90:128–37.

[51] Li W, Vanfretti L, Chompoobutrgool Y. Development and implementation of hydro turbine and governor models in a free and open source software package. Simulat Model Pract Theor 2012;24:84–102.

[52] Gencoglu C. Assessment of the effect of hydroelectric power plants governor settings on low frequency inter-area oscillations. Ms Thesis. Middle East Technical University; 2010.

[53] Sattouf M. Simulation model of hydro power plant using Matlab/Simulink. 6th EUROSIM Congress Model Simul 2014;4(1):295–301.

[54] Bhoi R, Ali SM. Simulation for speed control of the small hydro power plant using PID controllers. Int J Adv Res Electr Instrum Eng 2014;3(4):8392–9.

[55] Nassar I. Improvements of primary and secondary control of the Turkish power system for interconnection with the European system. PhD thesis submitted to Faculty of Computer Science and Electrical Engineering. Turkey: Rostock University; 2010.

[56] anskas LG, Baranauskas A, Gamage KAA, Azubalis M. Hybrid wind power balance control strategy using thermal power, hydro power and flow batteries. Electr Power Energy Syst 2016;74:310–21.

[57] Lone RA. Modeling and analysis of canal type small hydro power plant and performance enhancement using PID controller. IOSR-JEEE 2013;6:06–14.

[58] Ramey D, Skooglund J. Detailed hydrogovernor representation for system stability studies, power apparatus and systems. IEEE Trans PAS 1970;89:106–12.

[59] Sattouf M. Data acquisition and control system of hydroelectric power plant using internet techniques. (Doctoral Thesis). Retrieved From: Brno University Of Technology.

[60] Gbadamosi SL, Adedayo OO. Dynamic modeling and simulation of Shiroro hydropower plant in Nigeria using Matlab/Simulink. Int J Sci Eng Res 2015;6.

[61] IEEE Standards Board. IEEE recommended practice for excitation system models for power system stability studies. IEEE Std 421.5–1992.

[62] S. Mcfadyen. Photovoltaic (PV) – Electrical Calculations. Available at: http://myelectrical.com/notes/entryid/225/photovoltaic-pv-electrical-calculations.

[63] Tan R, Chow T. A comparative study of feed in tarrif and net metering for UCSI University North Wing Campus with 100 kW solar photovoltaic system. Energy Procedia 2016;100:86–91.

[64] Tan R. PV feed-in-tariff and net metering model, MATLAB central file exchange. 2016.

[65] MATLAB/Simulink release. Natick, Massachusetts, United States: The MathWorks, Inc.; 2012b.

[66] Jelali M, Kroll A. Hydraulic servo-systems: modeling, identification and control. London: Springer-Verlag London Limited; 2004.

[67] Totten G. Handbook of hydraulic fluid technology. New York, NY: Marcel Dekker, Inc; 2000.

[68] Ilango S, Soundararajan V. Introduction to hydraulics and pneumatics. 2nd ed. New Delhi: PHI Learning Private Limited; 2012.

[69] Olsson H, Astrom KJ, de Wit CC, Gafvert M, Lischinsky P. Friction models and friction compensation. European J Control 1998;4(3).

[70] Samyn P, Quintelier J, Ost W, De Baets P, Schoukens G. Sliding behavior of pure polyester and polyester-PTFE filled bulk composites in overload conditions. Polym Test 2005;21(5).588–603.

[71] Kim HK et al. Approximation of contact stress for a compressed and laterally one side restrained O-ring.

[72] Yokoyama K, et al. Effect of contact pressure and thermal degradation on the sealability of O-ring. JSAE Rev July 4, 1997;19(2):123–8.

[73] Johnson KL. Contact mechanics. Cambridge: Cambridge University Press.

[74] Imnoeng. Moody friction factor calculation. LMNO Engineering. Research, and Software, Ltd.; 2014. Retrived from, http://www.lmnoeng.com/moody.php.

[75] Larock B, Jeppson R, Watters G. Hydraulics of pipeline systems. 1st ed. Boca Raton, FL: CRC Press; 2000

[76] Aufleger M, Neisch V, Robert Klar R, Lumassegger SA. Comprehensive hydraulic gravity energy storage system eboth for offshore and onshore applications. In: E-proceedings of the 36th IAHR World Congress (Netherlands); 2015.

[77] Flitney R. In: Flitney R, editor. Seals and sealing handbook. 5th ed. Oxford: Elsevier Science; 2007.

CHAPTER 6

Energy Storage Challenges

ASMAE BERRADA, PHD [1] • SALMA BOUJMIRAZ [2]

[1] *School of Renewable Energy, LERMA Université International de Rabat;* [2] *School of Science and Engineering, Al Akhawayn University*

INTRODUCTION

An increasing interest in the use of energy storage (ES) as part of the electricity grid operation has been witnessed recently due to the evolvement and growth of the energy industry. The two important developments occurring within the energy industry include the high integration of renewable energy and the introduction of restructured markets. As a result of this flourishment, more research in the field gave birth to various novel storage technologies. As discussed in Chapter 1, there are currently a number of energy storage systems (ESSs) that have been successfully developed such as pumped hydro storage (PHS) system, thermal system, and compressed air energy storage system (CAESS). These systems have been proven for their availability and reliability. Other storage systems, such as batteries, have seen great improvement and advancement in their field. There is no doubt that ESSs increase the integration of variable renewable energy sources and reduce greenhouse gas emissions. However, in spite of these progresses, ES technologies are still facing significant challenges in the whole world.

The development of ES technologies is significantly linked to innovation. Nonetheless, this latter is more likely to be affected by the severe competition between ESSs and conventional generators, which is mainly due to the current immature regulatory regime. At present time, ES high capital investment coupled with regulatory barriers make its competitiveness very weak in most electricity markets. A number of countries are considering regulatory improvements to encourage ES use.

ES provides various benefits to the power system as it increases both its reliability and economics. Some of the services provided by ESSs are remunerated by markets; whereas others are provided to customers. To assess the value of ES, several factors need to be taken into consideration such as the technology-specific characteristics, location, type, and time span of deployment. The compensation of regulated services is acquired from charging utility customers, whereas the compensation

of market-remunerated services is acquired from competitive markets revenues.

In this chapter, different aspects related to ES opportunities and barriers will be studied. Case studies about ES deployment worldwide will be discussed in section Energy Storage Worldwide. Then, energy markets will be introduced followed by an overview of ES drivers, application, and value streams. Finally, the different barriers affecting ES deployment will be investigated in detail.

ENERGY STORAGE WORLDWIDE

The modern society and the global economy and security depend heavily on the electric grid. This latter has been formed in the 20th century and is permanently undergoing transformations. The use of novel energy resources is rapidly growing. Renewable energy sources (RES) are a reliable way for energy generation especially for countries going through a fast development. They represent an attractive competitive and sustainable alternative to the conventional energy sources as there is an abundance of natural resources. The market penetration of RES around the world is becoming more and more common [1]. The popularity of these types of energy sources might come with some drawbacks as their main limitation is their random, variable, and uncontrollable energy output. This can affect the grid management bringing some stability-related issues. It is necessary to invest in ES to ensure a resilient, safe, and reliable energy supply, as well as balance the energy demand and supply.

ES has a great potential to support the undergoing transformation of the grid. The integration of ES as part of the electricity system is not a new concept. In fact, it existed before the introduction smart grid systems. Pumped hydroelectric storage (PHES) technology has been deployed in bulk ES accounting for almost 99% of ES in the world [2]. Nonetheless, the construction of this technology requires not only suitable sites but also extremely large surfaces and an important capital investment. For this reason, grid operators are

Gravity Energy Storage. https://doi.org/10.1016/B978-0-12-816717-5.00006-2

looking for alternative ES technologies able to store electric energy and discharge it at a different instant with controllable power modulation and time delays.

It is necessary to establish a suitable support structure to allow alternative ESSs to be widely deployed. Thus, the main goals of this chapter are to explore the use of ES technologies in different countries as well as provide an insight about its diverse barriers.

Storing Energy in Japan

Japan is one of the world's leading countries in renewable energy markets, smart-grid, and ES technologies. ES is an important element contributing to energy independency and grid stability of the country. Japan is an isolated island, which hinders its connection with neighboring countries for energy supply when needed. ESS contributes to about 15% of the country installed generation capacity [3]. The need of investing in ES has been increased due to the Fukushima disaster, which resulted in a loss of public trust in nuclear energy. As a result, the Noda administration revised the traditional national energy policy to encourage the use of RES. Before the Fukushima crisis, RES accounted only for 3% of the total generation capacity since 1973. After

the tragedy, at the end of September 2012, the new energy policy of the country is supposed to reach 20% of electricity generated within 10 years from RES and 30%—35% by 2035 [4]. With the increasing deployment of RES in Japan, it is important to invest in ES not only to ensure energy security but also to maintain grid stability and resilience.

Today, ES landscape is widely distributed in Japan as shown in Fig. 6.1. The map data are taken from the DOE Global Energy Storage Database [5]. The ES infrastructure is interconnected with smart-grid as well as smart-city infrastructure. This map presents large-scale ES sites only, which include municipal-, utility-, and industrial-scale ES. The sites that are shown in these figures have a capacity greater than or equal to 300 kW [5]. PHES is concentrated in the mountainous and central region of the country, whereas battery-based ESSs are sited in highly populated and east coastal regions. Japan is a leading country in battery ES, particularly sodium-sulfur (NaS) battery storage. Worldwide, this market is dominated by NGK, a Japanese ceramic company, in cooperation with Tokyo Electric Power Company Holdings (TEPCO) that entered the research field and started commercializing batteries in 2002 [6].

FIG. 6.1 Japan energy storage landscape map.

TABLE 6.1
Transition and Forecast of Japan's Total Stationary Energy Storage System Market Size and Annual Growth Rate (2012–20) Excluding Pumped Hydroelectric Storage.

	2012	2013	2014	2015 (Prospect)	2016 (Forecast)	2017 (Forecast)	2020 (Forecast)
ES market size (kWh)	109,346	189,189	299,643	581,491	634,252	1,195,708	3,306,600
Year-on-year growth (Y-o-Y)	—	173.0%	158.4%	194.1%	109.1%	188.5%	276.5%

The Japanese public sector supported ES since the 1970s oil crisis. In 1978, research about ES, heat pump, and gas turbine technologies was encouraged through a 10-year sponsored program by the Ministry of Economy, Trade, and Industry (METI) R&D [7]. Moreover, in 1980, the New Energy and Industrial Technology Development Organization (NEDO) was established in response to the need of new ES technologies to remediate the long construction periods and large budget of large-scale PHES facilities. This organization has launched new battery technology development during the 10 following years. The main goal of this research project was to reduce Japan dependency on oil imports through encouraging the development of new efficient energy conversion technologies. In fact, research has not only led to the development of NaS batteries but has also resulted in the improvement of vanadium redox flow battery (VRFB), lead-acid (Pb-acid), and zinc bromide technologies. Many other programs were established aiming to continue funding important R&D in this field.

In the 21st century, the Japanese public sector continued supporting ES through the launching of numerous programs. The NEDO started a technology development project in 2006 that consists of the development of high-efficiency, low-cost, and long-lasting ES technologies. Furthermore, in 2012, METI initiated a storage battery strategy project whose main goal is to develop and apply plans and policies related to battery storage and its market [8]. In 2014, Japan's fourth strategic energy plan was launched. Its target was to reach a share of 50% of the world's 2020 forecasted global storage market, which was estimated for ¥20 trillion [9].

Concerning the current legal and regulatory infrastructure of Japan, the Electricity Business Act and the Fire Prevention Ordinance made distinct difference between large-scale and small-scale ES usage. The regulation and norms are more focused on large-scale ESSs as the safety and energy supply risks related to this type of storage are higher. Small-scale storage is privileged by the regulatory structure as it has lower regulatory obstacles compared with large-scale ES. By 2020, the current target of ES market is to have a share of 25% of the residential and industrial global battery storage market and a 35% of the global large-scale battery storage [10].

Japan's stationary ESS market size is growing massively and is projected to increase over the next years. Table 6.1 presents the size evolution of the Japanese ESS market excluding PHS. This is mainly because this technology has reached its economic maturity and is not considered compatible with small-scale ES markets in which most of the research is oriented [11].

Storing Energy in the United States

The United States pursuit of a sustainable and clean future has led to the proliferation of the efforts related to the development of ES. Before 2010, PHES was the only available storage technology for grid-scale storage with nearly 20-GW capacity in the United States. Grid-connected ES started to grow after 2010 because of the increasing penetration of RES and the presence of diverse constraints related to building new PHES [12]. In 2018, ESSs represented about 2.5% of the total electric production capacity of the country, which is around 25.2 GW and knowing that the share of PHES in this country is around 94% [5]. California is the leading state in ES with a capacity of 4.2 GW and more than 220 operational stations [13]. The existing and functional ESSs in the United States are classified by the DOE Global Energy Storage Database into four main categories, which include electrochemical, electromechanical, PHS, and thermal storage [5]. As of October 2018, the installed capacities of the aforementioned technologies are listed in Table 6.2. The Energy Storage Association (ESA) has decided to adopt novel, advanced ESS by the 2025 reaching a total capacity of 35 GW or more. There are numerous ongoing research and works

TABLE 6.2
US Operational Energy Storage Systems and Their Total Capacities (2018).

Technology Type	Total Number of Operational Projects	Total Capacity (MW)	Number of Storage Systems With a Rated Power > 10 MW
Electrochemical storage	338	764	27
Electromechanical storage	26	172	3
Pumped hydrostorage	38	22,549	37
Thermal storage	155	665	5

related to the development of ESS in terms of not only efficiency but also capacity and design.

In the United States, ESS and their services can be used in both regulated and deregulated markets. Nonetheless, there are well-established regulations and rules that provide the business case and economics framework of electric-grid ES-connected technologies. The supervision of these rules and regulations is taken care of and governed by different institution and entities, such as Federal Energy Regulatory Commission (FERC), Public Utility Commissions (PUCs), and the Independent System Operators (ISOs) in some regions [14]. This supervision can have an effect on the storage industry growth as some policies can either encourage or hinder market opportunities. The codes, standards, and regulations (CSRs) related to ESS in the United States are crucial in maintaining a safe and reliable use of the technologies [15]. CSRs encapsulate all ES aspects ranging from the fundamental elements to the operational ones.

The development of ES in the United States is encouraged through different programs. The US DOE and its Office of Electric Delivery and Energy Reliability are important institutions playing a crucial role in encouraging the development of ESSs. They have started the "Energy Storage Systems" program with the aim to conduct fundamental research and provide knowledge and financial-based supports for the use of ES technologies. The program has also a long-term goal that targets the deployment of new inexpensive technologies and at the same time decline the use of less promising ones. The DOE has developed another type of program called "Energy Innovation Hubs" aiming to influence the industry, academic institutions, and the government. The hubs are intended to promote the utilization of ES technologies and encourage their advance and commercialization. The promising technologies are supported by the partners of the program, facilitating and boosting their market penetration [16].

In addition to the diverse programs initiated by DOE, the Advance Research Projects Agency-Energy has also been established in 2007. This latter was created as a consequence of the decline faced by the country in the area of science and technology. In 2009, $400 million were allocated to the first project of the agency under the president Barack H. Obama and supported by the American Recovery and Reinvestment Act in 2009. It has also received its first congressional funds in 2011and 2012 [17]. A portion of these funds was allocated for ES development. In 2018, as part of the ARPA-E program, the US DOE has declared the selection of 10 new projects, as shown in Table 6.3, with the aim of extending grid ES [18,19]. This program, entitled Duration Addition to Electricity Storage (DAYS), enables awardees to improve the performance and resilience of the grid through the development of long-duration storage systems that can be used in any location.

In 2010, California PUC had been required to implement a storage capacity of 1.33 GW by 2024 [15]. Moreover, in 2013, Order No. 784 had been issued by the FERC targeting public utilities for better ES devices deployment [16]. These regulations have a direct impact on customers as they have access to better and less-expensive ancillary services. The biggest concern related to code standards and regulations is their inability to keep up with the fast development of ESS.

Storing Energy in Germany

In Germany, the ES installed capacity is 7 GW representing 8% of the total wind and solar installed capacity [5]. According to the DOE, the country has about 84 ES systems, which is equivalent to 7543 MW. There exist five major types of operational technologies in Germany, which are electrochemical storage, electromechanical storage, hydrogen storage, PHS, and thermal storage. As shown in Table 6.4, the technology with the highest capacity after PHS is electromechanical storage with a

TABLE 6.3
List of DAYS Awardees Project and Their Budget.

Awardee	Project Title	Budget
National Renewable Energy Laboratory	Economic Long-duration Electricity Storage by Using Low-cost Thermal Energy Storage and High-efficiency Power Cycle	$2,791,595
Michigan State University	Scalable Thermochemical Option for Renewable Energy Storage (STORES)	$2,000,000
Brayton Energy, LLC	Improved Laughlin-Brayton Cycle Energy Storage	$1,994,005
Form Energy, Inc.	Aqueous Sulfur Systems for Long-duration Grid Storage	$3,948,667
Quidnet Energy, Inc.	Geo-mechanical Pumped Storage	$3,298,786
Primus Power	Minimal Overhead Storage Technology for Duration Addition to Electricity Storage	$3,500,000
University of Tennessee, Knoxville	Reversible Fuel Cells for Long-duration Storage	$1,500,000
Echogen Power Systems (DE), Inc.	Low-cost, Long-duration Electrical Energy Storage Using a CO_2-based Pumped Thermal Energy Storage System	$3,000,000
United Technologies Research Center	High-performance Flow Battery with Inexpensive Inorganic Reactants	$3,000,000
Antora Energy	Solid State Thermal Battery	$3,000,000

TABLE 6.4
German Operational Storage System and Their Total Capacity (2018).

Technology Type	Total Number of Operational Projects	Total Capacity (MW)
Electrochemical storage	46	320
Electromechanical storage	4	877
Hydrogen storage	7	17
Pumped hydrostorage	29	6688
Thermal storage	1	2

capacity of 877 MW. This includes compressed air storage and flywheel. There are two compressed air storage projects with a combined capacity of 490 MW and two flywheel projects with a total capacity of 387 MW.

ES is considered an essential element in the energy transition of the country. To reach its goals, Germany needs to transform its power system and market and should establish new energy policies. The Energiewende is composed of two fundamental elements, which are to

- achieve the nuclear phase out by 2020 and
- increase the integration of RES.

Germany has set a target of generating electricity from renewable sources reaching a minimum of 35%, 50%, and 80% of the gross power consumption by 2020, 2030, and 2050, respectively [20]. The country will invest mostly in wind and solar energy to achieve its goals mainly because it has a limited potential of hydropower and biomass sources.

Germany has encouraged research related to ES technologies in many ways not only at the ministries level but also at the state level. For instance, it has provided diverse types of funds to research institutions and universities to support their activities. The most important funding initiative is the "Förderinitiative Energiespeicher." This initiative is part of sixth Energy Research Program and is organized by the three German ministries, which are the BMWi, BMU, and BMBF. It is targeting all types of ESSs with no restrictions as long as they are stationary and beneficial for the German energy system. Table 6.5 shows some of the most important projects funded by the BMU ministry [21].

The market-pull programs in Germany provide a different type of support to improve the country's ES situation through facilitating the investment in this field. KfW is a state-owned bank that plays a huge role in the development of Germany; it assists individuals, companies, cities, or municipalities through diverse

TABLE 6.5
Top Funded Storage-Related Research Projects by the BMU.

Beneficiary	Project Description	Fund (Euros)
Zentrum für Sonnenenergie-und Wasserstoff-Forschung Baden-Württemberg (ZSW)	The power to gas project consists of building and operating a storage research facility of electricity as methane (250 KWe).	3,500,000
Helmut Schmidt Universität Hamburg	Development of a measurement device enabling the determination of the frequency-dependent line impedance on the high-voltage level.	2,400,000
FH Aachen	HiTExStor II project is about developing, demonstrating and testing a high-temperature moving bed heat exchanger used for storing sensitive heat.	2,100,000
Steca Elektronik GmbH	Developing photovoltaic island system using lithium-ion batteries to provide long-lasting storage.	1,500,000

TABLE 6.6
Operational Storage System and Their Total Capacity in China (2018).

Technology Type	Total Number of Operational Projects	Total Capacity of Operational Projects (MW)
Electrochemical	59	53
Electromechanical	2	1
Pumped hydro	34	31,999
Thermal storage	2	12

Storing Energy in China

As of October 2018, China has installed a capacity of 32 GW of operational ES. It has the second largest installed PHES capacity in the world following Japan and followed by the United States. It also plans to develop this type of storage in the future as it is expected to reach a total capacity of 40 GW by 2020 [22]. Other ES types are not widely deployed as shown in Table 6.6.

China Energy Storage Alliance (CNESA) is the first ES association in the country. It was founded in 2010 with a mission to influence the government policy to incite and support the growth of RE through the use of reliable energy systems. China released its first national-level ES policy document in October 2017 aiming to guide the industry proliferation referred to as "Guiding Opinions" [23]. The document was structured around three main elements which include the following:

- Confirm and support the necessity of ES in the development of smart grids, RE grid-penetration, and the "Internet of Energy."
- Fix development goals for the forthcoming 10 years.
- List the required policies and guidelines to support and insure the achievement of the goals set.

This document states the importance of ES deployment to the resilience, stability, and reliability of the modern power grid. The objectives set in the document give a clear idea about the near future of the Chinese industry. To achieve technology advance, the policy document listed plans for diverse ES technology demonstrations, such as 100 MW-grade Li-ion battery demonstration. If this ESS were to be constructed, for instance, it will represent a great technology breakthrough since the largest Li-ion battery capacity in the world is 30 MW/120 MWh in California [23]. To achieve the goals of the "Guiding Opinions," it has also called subordinate government agencies for establishing new policies to regulate diverse aspects related

programs allowing them to get low interest loans as well as governmental subsidies. Besides financing a big range of business projects, KfW is particularly supporting projects related to renewable energy, energy efficiency, and ES. In the T&D market, this bank is supporting ES projects through two programs. On the one hand, project number 203 offers low-interest loans to the municipalities interested in extending current storage station and in constructing new ones. On the other hand, project 204 provides municipalities and medium-sized public-private partnerships with low-interest loans that can reach up to €50 million. There is also the PV battery storage subsidy program that supports decentralized battery storage for PV installation. The initialization of this policy is taken care of by KfW through program 275.

to ES, such as regulating pricing instruments and tools as well as technology standards.

After the national-level ES policy release, China has increased the drafting as well as the release of new regional energy policies [24]. This has led to the concretization of diverse ES projects, which had a direct impact on the development of the operational ES capacity in the country. In fact, in 2018, it has attained a growth of 281% higher than the entire previous year. Provinces in China differ in a variety of ways in terms of power structure and consumption, resources distribution, electricity pricing, etc. All these divergences had influenced the way ES projects were implemented in each province. Nonetheless, electrochemical storage is the type of ES that was mostly used in all provinces, which includes lithium-ion batteries, lead-carbon batteries, as well as vanadium and zinc bromine flow batteries. Henan and Jiangsu were the first provinces in the country to release grid-side level projects of 100 MW. These projects give hope for the use of large-scale storage in the future by other provinces or even the country. Nonetheless, the ES benefits valuation is still not recognized in China. ES market mechanisms are far behind compared with other countries.

Storing Energy in Morocco

ES represents a great opportunity for investors as well as the power system in Morocco. The total installed ES capacity is around 1.3 GW with a total of six projects as shown in Table 6.7. Only four projects are operational with a capacity of 826 MW. The oldest ES station in Morocco is the Afourer Pumped Storage Scheme. The construction of this station started in 2001 and ended in 2004. The project was funded by Arab Fund for Economic & Social Development. Ait Baha plant thermal storage operation started in 2014 and was developed by Airlight Energy in the region of Agadir. In addition, Noor 1 CSP solar plant in Ouarzazat deploys molten salt ES with a storage duration of 3 hours. Similarly, Noor 2 plant deploys the same storage technology as Noor 1, yet it has a storage duration capacity of 7 hours. Finally, Noor 3 is still under construction, and it is expected to use the same technology as the other Noor projects and have a storage capacity of 7 hours.

The Moroccan legislative framework does not separately define ES. The rules regarding the issue of ESSs are stated by the law applicable to the production of energy. However, the development of legislation on ES is expected in the near future because of the increasing penetration of renewable energy.

Specific recommendations for a "successful transition into a Green Economy" have been issued by the Economic and Social Council for Green Energy; among which the need to accomplish a competitive electricity market. In this regard, a number of recommendations are proposed [25]:

- Solving issues associated with ES and peak energy demand.
- Providing a varied mix of energy sources such as the use of biomass, clean coal, and PHS.
- Installing small and medium power stations; especially renewable energy installations.

Tasks that should be accomplished by the Ministry of Energy have been set out by the legislative decree n° 2-14-541. In the context of a liberalized energy market consolidation in Morocco, the Ministry of Energy has to develop a key ES policy and to control the

TABLE 6.7
List of Energy Storage Projects in Morocco.

Station Name	Technology Type	Power (kW)	Status
Afourer Pumped Storage Scheme	Pumped hydrostorage	465,000	Operational
Abdelmoumen Pumped Storage Power Station	Pumped hydrostorage	350,000	Announced
Ait Baha Plant Thermal Storage—Airlight Energy	Heat thermal storage	650	Operational
NOOR I (Ouarzazate) CSP Solar Plant	Molten salt thermal storage	160,000	Operational
NOOR II (Ouarzazate) CSP Solar Plant—ACWA	Molten salt thermal storage	200,000	Operational
NOOR III (Ouarzazate) CSP Solar Plant ACWA	Molten salt thermal storage	150,000	Under construction

functioning and organization of the energy markets as stated by Article 1.

Battery storage standards have been enacted recently by the Moroccan Institute for Standardization (IMA-NOR) [26]. These standards include NM CEI 61427-1, NM EN 12977-3, NM EN 12977-4, which regulate the general conditions, the performance testing methods applying to the storage installations for water solar heating, and the conditions applying to the combined storage methods for solar heating, respectively [27].

The conditions under which renewable energy systems can be installed and operated are specified by law 13-9 [28]. This law has liberalized the system of energy production based on RES. The development of law 58-157, which amends the law 13-09 has brought some interesting modifications [29]; such as increasing the threshold of electricity production from hydro, providing access to low-voltage electricity grid to electricity stations referred to in law 13-09, and allowing the selling of excess electricity production to the National Electricity Office (NEO) or to the energy transmission system operator. Therefore, future consequences related to ES in Morocco might results because of the development of these laws. However, the question of ES is currently not regulated.

ES is facing a number of challenges in Morocco as it is still at a development stage. These include the following:

- ES regulation barrier due to the lack of a specific legislation.
- Limitation of reselling electricity to ONE: Even though the law 58-15 permits the resale of excess energy production to ONEE or to the energy transmission operator; a restriction of 20% of the produced energy is applied in the resale.
- Variability of energy production: The production of energy from hydrosources in Morocco, for example, depends on pluviometry and varies significantly each year. The energy produced by hydropower stations in 2012 dropped by 15% in comparison with 2011 due to drought [30].
- Public regulation entities: Morocco does not currently have an energy transmission system operator. The legislation n° 48-15 emphasizes the need to establish an independent entity in charge of this role within ONE; along with the creation of an independent regulator. ONE is responsible for the production, the transmission and distribution (T&D) of energy. In addition, ONE is responsible for the production of renewable energy, in parallel with public and private entities. In this respect, ONE is the relevant entity in charge of ES projects. The council of government has implemented, in 2015, the

legislation n° 48-15 regulating the energy sector, and creating a National Authority for Electricity Regulation (ANRE) "Autorité Nationale de Régulation de l'Electricité". According to this draft, ANRE will be responsible for controlling the energy market.

ENERGY STORAGE MARKET

There are many factors contributing to the deployment of ES in each region. The ES potential of a country is influenced by a combination of diverse elements, such as its energy resources, electric grid infrastructure, electricity market, energy demand trends, legal and regulatory framework. The aforementioned factors define the need for new services and commodities. ES is a solution to satisfy these demands.

ES market is tightly linked to the physical structure of the electricity system. The power system architecture differs from a country to another. These differences affect the distribution circuit structure, which in turn affects the customer-sited ESS specifications. Thereby, it is crucial to understand the distribution network system in the realm of ES market development. In general, the distribution systems can be classified into two main categories, the North American distribution system and the European distribution system [31]. Both types share the same hardware elements, yet they differ in the layout and configuration.

In North America, the power distribution grid has a radial layout with moderately long feeder circuits, with many step-down transformers per feeder. Only few clients are served by each transformer. This type of power distribution network can be seen in other countries such as Latin American ones, Australia, and New Zealand. Compared with North America, Europe has denser populated region distribution. As a consequence, its distribution circuits' length is shorter with a smaller number of step-down transformers per feeder and more clients per transformer. Outside North America and Europe, these systems are used in different parts of the world, yet European practices are more widely deployed [31]. Hybrid systems, a mixing between the two systems, are also used in some regions especially the developing ones.

The regulatory framework and market structure are very important factors determining the electric power industry competitiveness as well as the potential of ES use. There exist two types of market models for national grids, which include the fully regulated markets and the deregulated ones. As for the regulated markets, one single entity is responsible for controlling electricity generation, distribution, and sale. In contrast, in deregulated markets, this is more liberalized, and many competitors exist in both the generation and retail level; clients are free to choose their own supplier. Nonetheless, many countries tend to use a hybridized model. This leads to the emergence of more complications related to the regulation framework of the whole system. The market structure allows the determination of the ESS final customer in the market as well as the ownership model and service performed.

ES development is partially influenced by energy usage trends of a given population. The energy usage patterns allow the determination of the power grid structure. Countries with crowded urban areas will require concentrated circuits that are able to deliver high voltage, such as the European model circuit. The urban area growth rate has also to be taken into consideration. A fast urbanization or high urban area growth rate implies the need to invest in the development of the electric infrastructure, which is theoretically a driver to create an ES market. Isolated communities depending on remote power systems fueled by a fossil fuel are the first populations to adopt and deploy ES. This is mainly because it allows them to reduce the consumption of the fuel, which makes ES profitable. Both a community population mix and its growth rate are important factors contributing to the definition of the type and size of the ES market.

The potential market for stationary ESS is also determined by the electric grid stability of the country. The size, type, and deployment of the ESS will be primarily influenced by the grid stability. For instance, distributed energy storage systems (DESS) will be more popular in areas where the grid is somehow unstable. In that case, it will be mainly used to prevent electric interruptions or blackouts. DESS will also be beneficial for industries where a minor electric issues or disturbances can result into severe damages. Moreover, unstable grids operators are more prone to use large-scale ES to minimize the risks of energy outages that could affect an important number of their customers. In some cases, the age of the grid infrastructure influences the ES demand, especially in developing countries aiming to attract more industrial investors. Investing in ES is becoming a more economically viable alternative compared with replacing the infrastructure of the grid since its investment cost is declining with time. With all of the aforementioned reasons, ES will become a more attractive alternative to conventional infrastructure upgrade.

Energy Storage Drivers

Every particular country or region has its own ES market trend. However, at a global level, ESS is mainly used for improving grid performance by enhancing its efficiency, resilience, security, sustainability, and at the same time providing additional services. The market segments of stationary ESSs can be divided into three main categories:

- Utility-scale systems
- Behind-the-meter systems
- Remote power systems

The utility-scale system is a large-scale storage system that provides services to the operators of the grid. This is becoming more common with the growing rate of renewable energy use. In fact, the key driver for utility-scale ES deployment is the increasing development of RES in the world. The power generation output of this latter is not suitable for conventional grids. ESSs are used for smoothening up the variable power output of renewable energies. They also help minimize the curtailments and enable a better supply and demand alignment. Another driver for the development for this type of ES is the global governmental effort to lower carbon emissions. For example, in 2015,197 countries approved to reduce their emissions to limit global warming at the Paris Agreement. To achieve their goals and targets, many countries have set their own targets to mitigate their emissions. These types of global agreements coupled with the falling price of renewable energy generation are threatening the competitiveness of the fossil fuel—powered power plants. The replacement of powered plants by renewable source is becoming more and more common. This has given rise to the need of integrating ES into the system to remediate stability issues. Utility-scale storage deployment is also driven by the necessity to modernize and develop the infrastructure of the grid. The rapidly growing population in developing economies with old grid infrastructures creates the need to invest in modern technologies. ES is essential in this particular type of investment as it plays an important role in the protection of the grid from different threats.

Behind-the-meter systems are technologies operating at the client side of the utility meter; they are the most commercialized ESSs. These systems are very attractive not only to customers but also to utilities as they work as a backup in case of a grid failure to provide electricity. Accurate sizing needs to be taken into consideration for the system to work appropriately. If the system is too small, it will not be able to support the loads when needed. On the other hand, if it is oversized, all the critical loads will be supplied, yet the system will be excessively expensive. Backup power is in fact the strongest advantage this system can provide.

The most important driver of behind-the-meter ES is the ability to enable customers to shift the load curve to avoid peak hours and hence reduce tremendously the electricity bill. In emerging markets, the main value provided by ESS is its use in storing energy for distributed generation (DG), such as photovoltaic (PV). In the majority of developed countries, the use of PV is encouraged by the government through enabling the customers to feed the excess of energy to the grid and get paid for it. Nonetheless, in Organization for Economic Co-operation and Development countries, this type of compensation was either eliminated or replaced by another type because of the prosperous development PV installations have known. In the case of compensation programs, elimination or inexistence, as well as when the compensation rate is lower than the retail price, ESS use becomes an economically viable and beneficial alternative. Customers can use the energy stored in the battery in peak demand periods when electricity is the most expensive in dynamic pricing markets.

Behind-the-meter ES has a great impact on distributed renewables deployment, mainly PVs. Many problems can be caused due to high feedback power by DG systems to the distribution system of the grid. The majority of the latter design in the world is not suitable for high electricity feedback. This issue can be remediated either by modernization of the system equipment or by setting feedback boundaries and controls. Behind-the-meter storage helps prevent this problem by storing energy on the customer side. The increase of distributed energy resources use in the world will affect differently every region or country. The impact will mostly depend on the existing centralized generation system of the area as well as the intended modifications to the system. Many countries are planning to deploy more DG systems in the upcoming decade coupled with their existing centralized systems. In some of them, the capacity of DG will superpass the centralized one. This will lead to the imperative need to restructure the energy market as well as the physical structure of the electricity network.

The third type of stationary ESSs is the remote power systems. This type refers to power systems that are used in isolated energy networks. These networks can be differentiated primarily by their size. The large remote systems are called microgrids and are composed of smaller units referred to as nanogrids. Nanogrids have shown that the "larger is better" economy of scale, which has reigned during the last century in the energy market, is not valid all the time. Sometimes, the use of this system by itself can have greater advantage than the microgrid type. Another criterion of differentiation between these systems is their proximity to the centralized grid. Tied to grid systems, refer to the systems directly linked to the grid. On the contrast, remote systems are autonomous and not dependent on the centralized grid. The most famous remote systems in the world are the diesel-fueled ones. The main driver of this market is shifting the use of high-cost and polluting diesel to the use of RES. If the shift occurs, a high need to invest in ES coupled with novel control strategies will arise to make operation more reliable.

The advantages provided by ESSs to remote systems are very similar to the ones provided to the conventional grid. It is important to understand the benefits of the services provided by these systems to accurately estimate their business viability. This applies particularly to developing countries in which fossil fuels are cheaper. However, building or modernizing the conventional centralized T&D grid infrastructure to cover the increasing electricity demand in these countries is extremely expensive, which can even be costlier than building new remote power systems.

Energy Storage Applications and Value Streams

ESSs provide a vast range of benefits to the traditional centralized energy generation system. ES can offer services that are not provided by conventional systems. As mentioned in Chapter 4, ES can be used to provide energy arbitrage; that is storing energy when electricity market prices are low and sell it when the price is higher. It can also provide similar services offered by traditional generation systems such as ramping capabilities. Existing power plants are not all equipped for rapid ramping or frequent cycling. However, most ES technologies are designed for that purpose. The integration of variable energy generation into the traditional power system requires additional features to preserve grid stability. ESSs are able to provide services not only to the generation side but also to the load side to maintain system stability.

It is important to know that not all ESSs are able to offer all the services discussed in this chapter. These services are not only site specific but also technology specific. Nonetheless, many services can be provided for a particular application. At the transmission level, for instance, storage offers both energy arbitrage and ancillary services. It can instantaneously mitigate transmission system congestion and contribute in meeting peak loads. At the customer-side level, ES can reduce the client peak demand and costs as well as improve power quality.

TABLE 6.8
Energy Storage Applications.

Application	Description
Price arbitrage	Storage permits electricity to be imported/stored when there is an abundance of electricity or when it is cheaper. It can also be exported to the grid when it is needed or when there is a rise in the electricity prices.
Ancillary services	Storage can allow storage providers to generate a variety of revenues from providing ancillary services that can range from frequency control to voltage response.
Demand reduction and peak shaving	Storage devices can be used by energy consumers to reduce their demand in peak times and thereby have control on their electricity bills.
Grid reinforcement and deferral	Storage devices can provide reinforcement for the grid through reducing peak demand as well as generation flows. This process is particularly provided during peak demand periods, which are usually very prompt or infrequent.
Variable generation integration	Storage allows an efficient use of variable generation when used along with renewable energy generation. It can store energy depending on the need, which enables generators to optimize the price of electricity and maximize its export.

There has been recently an undeniable burgeoning interest in ES, thanks to the development of the electricity industry. However, it is still a field that is facing a number of barriers, which hinder its large-scale deployment. Obstacles related to technical issues are often addressed, such as the cost and efficiency of technology; yet other nontechnical and policy-related matters seem to be side-lined. Before the introduction of restricted markets, the focus on ES use in vertically integrated utilities was solely to cover the energy needs during peak time. Nonetheless, currently, services offered by ES can go beyond that. Recent studies have proven its utility and ability not only to provide energy wholesale and capacity for large power systems but also to backup energy for a single home or building. ES can be used in a large range of applications. Examples of services provided by ES are listed and described in Table 6.8.

Flexible capacity value

Meeting peak demand periods is one of the most important drivers resulting in the addition of new energy generation resources. Because peak demand does not occur throughout the whole year, it is not economically profitable to invest in additional generation systems. ES is a good alternative to meet peak demand periods needs since they are relatively occasional. This service alone is not enough to justify the deployment of the expensive ES technology. However, the ability of ES to provide other services such as the delivery of flexible capacity makes it a great candidate. For

example, ESS can generate energy in a short notice and is also capable of efficiently operating over a large range of output levels. Flexible capacity is also advantageous to the customers. Behind-the-meter systems can mitigate the charges related to peak demand. It can be economically beneficial for the customers when value streams are combined such us backing up power and providing high power quality.

Arbitrage value

As explained before, energy arbitrage is the practice of storing energy during off-peak times and consuming it in peak times. The most common ESS used in power systems is pumped storage as it is able to deliver an enormous quantity of energy in the time needed. Capacity value coupled with energy arbitrage has been used for a long period of time as a justification to invest in ES. Nowadays, the prices in wholesale electricity market are relatively low because of the high solar and wind generation penetration. This can lead to negative market prices, which will have an impact on how energy arbitrage is perceived.

Renewable energy variability is likely to affect the wholesale markets by increasing the electricity price volatility. In other words, the prices would be tightly linked to the RES generation. When the generation is high, the electricity prices would drop and vice versa. This volatility gives the opportunity to an increase in the deployment of ES. However, this consequence is constrained by low natural gas prices as well as the abundance of generating ability.

Ancillary services and system balancing

The power system reliable operation relies on effective system balancing and ancillary services. In fact, energy transmission requires specific supports to transport energy from the generation to the loads and at the same time maintain reliable and effective operation. ES technologies can provide a large number of ancillary services that include system balancing, contingency reserves, reactive support services and voltage control, etc. These services need to be taken into consideration while assessing the economic benefit of ES. There is one important factor to take into consideration in this evaluation, which is to the extent to which there will be a possibility of coexistence or overlap of services, the degree of this existence within a specific market and the advance of specific existing valuing methods.

Power system balancing services refer to services that enable balancing between energy demand and supply while the power supply security and quality. For this purpose, generation reserves should be able to increase or decrease the generation of energy when notified under an automated control system. System balancing services are classified based on the period of time necessary for the device to respond, and they include regulation and frequency control, imbalance, and load-following reserves. Regulation reserves are considered the fastest responding service. Their respond time is few seconds when the system is under automated control. The second fastest service is following load reserves with a respond time of approximately 10 min. Finally, the slowest one is the imbalance reserves that are characterized with a ramp time ranging between 10 and 30 min.

The ramping ability is not a separate ancillary service itself; however, it is considered a generation characteristic. It refers to the rate at which the generation levels changes over time, and it is usually expressed in MW/mn. The day-ahead market creates hourly schedules to accommodate interhour ramps. On the other hand, intrahourly ramps can be covered by quick-start resources, which are usually combustion turbines. Renewable resource output is fluctuating rapidly and needs to be compassionate by resources with similar generation rates. The need of ramping is becoming more and more essential. Assessing the necessity of ramping is at its early stages. For this reason, there is a lack of ramping capacity standardization in the industry.

As mentioned in Chapter 4, contingency reserve services are necessary to maintain the required generation. They balance the loss of generation, which may occur because of a failure in the power system. Contingency reserve services can be classified into three main categories, which include spinning reserves, nonspinning reserves, and supplemental reserves.

Reactive support service is important to maintain a good electric power, and it is the product of current and voltage. In the alternating current power systems, the current and voltage alternate from positive values to negative values with respect to time. The voltage and current need to be synchronized to attain the optimum power derivation. The performance of the system depends on the synchronization of these two parameters. Issues related to voltage are usually related to poor reactive power support. Reactive support is the ability of special equipment, such as loads, generators, or other devices to correct the phase difference. The advantage of some ES technologies deployment is their availability and ability to provide reactive power support to the grid.

Some ES technologies can contribute to network stability by countering the disturbance effect within first few seconds. The latter is very important when a disturbance occurs, and the system needs to get back to stable operation. Automated tools can be used to improve the stability of the system. The responding time of services related to disturbances in the network is between the first 10 and 30 s. However, the support network is necessary to stabilize the system within the first 3 s.

Congestion management

ES technologies can be an ideal solution to remediate issues related to transmission congestion as mentioned in Chapter 4. Power generation is usually not located close to populated areas, which require high-voltage transmission lines for large distances. When the demand is high at peak periods, the transmission line might face congestion. This issue can be solved by putting the generation site closer to the populated areas. Nonetheless, there are many constraints related to placing the generation site next to a populated region such as air and nose pollution as well as land accessibility. Knowing that congestion periods do not last for a long period of time as they do not exceed few hours, investing in traditional generation next to cities would not be a feasible idea. This makes ES a good alternative to other energy generation types. There is another advantage to energy storage use for congestion management, which is the possibility to colocate it with other renewable generation. This creates an opportunity to better serve regions that are not well located for transmission facilities.

Besides the congestion occurring in the main high-voltage transmission regions, it can also occur locally in the distribution system when there is a high demand or high DG. In this case, small-scale storage can be a better option to avoid congestions compared with replacing some major mechanisms. The small-scale storage can either be installed at substations or even at the load site.

ENERGY STORAGE BARRIERS

Despite all of the aforementioned benefits, storage systems deployment is still limited. The economic aspect of ES is one of the main challenges faced by these systems. Several factors influence the economics of ES which complicate its evaluation. These include the system size, application, type, and location. In addition, ESSs are still very expensive. Their costs have to decrease in order to be attractive and economically viable. The capital cost of ES consists of costs of per unit of energy capacity ($/kWh) and per unit of power ($/kW). Some storage systems have low capital cost per unit of power but high cost of energy capacity such as flywheel and supercapacitors. Other systems such as PHS and CAES have low cost per unit of energy but high capital costs per unit of power. The storage costs include also the system operating cost, which consists of energy-related and non–energy-related operating cost. The energy-related costs are about the cost of energy used to charge the system and the cost to purchase power used to make up for the energy losses. The non–energy-related costs compromise the cost of labor associated with plant operation, maintenance cost, frequency of operating cycles, equipment depreciation cost, and all costs associated with decommissioning and disposal. The aforementioned costs vary widely from one storage system to another, in addition to the storage system size. For instance, operating large-scale CAES and PHS require costs associated with labor, while batteries used for small scale may not require these costs as they are designed to operate autonomously. In general, the most cost-effective systems used in for scale storage are CAES and PHS with frequent cycles. When the number of cycles is low, batteries are expected to be the cheapest option, whereas supercapacitors and flywheel are more ideal for frequent use and very short periods [32].

Another challenge faced by these technologies is associated with the system efficiency. The challenge lies in investigating and identifying methods to optimize ES efficiencies. In real-life applications, PHES have a round-trip efficiency of about 75%, although its theoretic prediction demonstrates that it can achieve an efficiency of 90%.

The third challenge is related to the lack of standard for the interconnections of ESSs with the electric grid. In addition, some storage systems still suffer from the difficult and complexity of being developed in a modular design. This later helps to promote the flexibility that the system provides. It enables for a better optimization of the system behavior in response to varying conditions. Batteries are an example of storage systems that are modularized and standardized. For a better prosper of ES, government support is highly needed.

The development of ESSs has lagged far behind the growth of intermittent RES. To accommodate the variability of these resources, several flexibility methods are being tapped more extensively. Emphasis is being given to cost-effective ways, such as demand side management and market efficiencies. However, with the high penetration of RES, power systems have to include other flexibility ways beyond what may be available today. This should be done in a reliable, economic, and environmental manner.

Competition on efficiency and price are considered the main barriers reasonably faced by developers. However, it is essential to analyze policies that obstruct the development of ES and investigate new policies required to overcome unnecessary obstacles to the cost-effective deployment of ES. Changes in policy should be considered in four major areas, which include ES valuation and markets, regulatory actions, system development risk, and standardization. Important efforts are currently underway to address some of the obstacles encountered by ES. Opening ancillary service markets to ESS are tackled by the FERC orders (755 and 784). Policies are being executed for implementing these orders to encourage the development of ES. These policies are considered a huge step forward for ES. They also represent a proof that this field still needs to be enhanced to guarantee an objective improvement of ES.

Benefits and Market Valuation

The most significant barrier facing the development of ES is the incapability of quantifying the multiple value streams provided by this latter to the electric grid [33]. The real value of ES is not based only on the difference between purchasing energy at low price and selling it at higher prices. It is rather associated with all the services they provide including ancillary services, regulation, load-following reserve, as well as T&D system support. ES brings other benefits such as low or zero emissions, rapid response, and bidirectionality of reserve capability.

Before the initiation of restructured markets in the United States, only two types of services were usually considered while valuating ES. These include load leveling and firm capacity. The valuation of other benefits such as ancillary services was rarely performed. This is mainly due to the limitations of capacity expansion and simulation software used by utilities in regulated markets. The advent of restructured markets and the initiation of ancillary service markets have resulted in the valuation of some services such as fast response services, which are provided by certain types of ESSs. Yet, the deployment of batteries and flywheels was mostly located in regions with restructured markets to perform frequency regulation reserve services. For example, the development of a 1-MW battery in PJM market, and a 3-MW flywheel project, as well as a 20-MW plant in the New York ISO market. These projects were made possible mostly by the FERC-issued order 890. This requires the consideration of nongeneration resources such as ES for electric grid services by wholesale markets. Consequently, several market operators have proposed new tariffs enabling ES to participate in ancillary markets. Policies associated with standard utility valuation should compromise other values brought by ES. For more than a decade, ancillary service markets have been in operation for a long time in California Independent System Operator (CAISO) and were generally irrelevant to total market value. However, because of the efforts placed on improving the quality of frequency regulation and response, they have attracted more attention.

Despite the development that ES has known as part of the electric grid operation, its market valuation is still hindered by existing constraints. This is mainly due to the lack of efficiency in price transparency for end users. The major ES applications serving the distribution network as well as the end user have very limited or even no exposure to the market. This exposure depends heavily on utility rate structures. Today, these structures are not able to capture the varying cost of electricity with time. Usually, those rate structures variation with respect to time can be predetermined, which make them not dynamically adjustable in real time. The lack of smart grid technologies in the electric industry limits the growth and deployment of real-time dispatch storage devices. This eliminates a great potential of economic participants in the market, but most importantly, it obstructs the achievement of other benefits. Managing energy costs through load shifting by end users is reasonably similar to energy arbitrage. In other words, it is not important if the device is sited at the load level or in the transmission network level. Currently, from the system complexity standpoint, it is much harder to plan and dispatch hundreds or even thousands of customer-sited storage devices rather than a smaller number of transmission-sited devices.

Customer-sited storage provides measurable and real benefits allowing the reduction of future T&D losses and infrastructure needs. The benefits of avoided T&D losses are several and can range from providing a more effective peak capacity of load-sited storage to a better T&D utilization. New infrastructures are, therefore, not necessary, thanks to all these advantages. Nonetheless, because the present distribution network is part of a regulated monopoly and the local marginal prices (LMPs) are not computed at the level of distribution network, it would be very unlikely to take advantage of these benefits without the intervention of some policies or the integration of smart grid technologies. Storage is not placed at the optimum position to take advantage of its fullest potential, and there are no incentives to achieve that. This challenge can be mostly distinguishable for ES technologies that can be load-sited only like thermal ES. It would be much easier to work on a novel business model that implement storage owned by the utility, yet that is customer-sited. This method is comparable with the intervention of third-party owners who act as a link between facilities and customers to facilitate the embracing of PV use.

As discussed in Chapter 4, storage can potentially provide a large number of services, and many of the values could be generated in existing restructured markets. An important issue with the valuation of ES is that most studies and analysis examine only one or two storage applications. The published works that focus mainly on energy arbitrage usually conclude that this service is unlikely to support the high investment costs of most ESSs. As a contrary, work of studies that include other services [34], such as providing ancillary services, demonstrates that among different ancillary services, regulation is considered the most valuable, followed by spinning and nonspinning reserves. Therefore, as previously noted, maximizing storage value will likely require multiple value streams.

Transmission limitations can be mitigated through incentive provided by restructured markets. This has been proved using historic data about LMPs to display the variation in arbitrage values within a transmission network. Storage advance could be encouraged by these revenue differences at the most congested areas in the network to get greater profits. The latter represents a storage development model that is very similar to a model supported by early market restructuring advocates, which is the merchant transmission model [35]. However, the presence of some facts related to electricity systems such as the market power and uneven investment might affect the investment efficiency, which will, in return, have a direct impact on storage investment.

The ability of storage to diminish the generator ramping and cycling is also an inadequately priced aspect of storage. These types of profits can be realized when ES provides energy arbitrage service. This benefit can become even more significant with the development of different renewables that increase the pressure on conventional generators. The cost of ramping is not taken into consideration as LMPs are normally calculated using a static optimal power flow, which does exclude generator ramping constraints. However, if it was to be computed using a dynamic model, the additional system costs enacted by the ramp will be appropriately allocated. Storage can also help reduce the costs of the startups of shutdowns of generators. Again, these costs are not signaled by the LMPs and consequently not reflected in the market. Therefore, the two stated storage benefits are clearly underestimated and devalued in the market mainly because of computational burdens and market design structure.

One limitation in the existing literature concerning the storage benefits valuation is the fragmentation of the studied applications that the storage offers. Despite the progress in research effort and results in examining the benefits of storage in diverse applications, the capability to quantify some of its values is still limited especially when the applications considered are combined. Storage benefits valuation is undervalued since a lot of studies consider only one or two closely related applications, whereas several value streams need to be considered to maximize its value. If other applications were to be examined and studied, storage might show in some cases an economically viable and favorable potential as some applications complement one another or compete with each other. Therefore, there values can be additional. This issue can be overcome by conducting more comprehensive studies on storage applications. In addition, adding the results of separate studies of different applications is not applicable. For instance, if storage is used for ancillary services, it can decrease its ability in the following hours to provide arbitrage. Some studies tried to evaluate multiple applications, yet they seem to neglect the complexity of these types of interactions between them. For example, Druri et al. studied the back-casting heuristic to determine the influence of the predicted assumptions on the arbitrage values, and in this case, storage use was optimized using historic price data [36]. They proved that this simple method could be very accurate. In addition, Mokrian and Stephen (2006) developed dynamic programming methods that help predict the optimal arbitrage value when the future price is uncertain [37]. These methods can be efficient in giving an idea about

storage revenues, yet they seem to fail in predicting accurately the net revenue of storage over its lifecycle. One more limitation of these approaches is that they are exclusively giving attention to the price uncertainty while other types of uncertainty might be more relevant.

Another limitation of the existing storage valuation approaches is that they seem to assume that ES operates in a flawlessly competitive manner by disregarding the strategic performance of the storage operator. Many arbitrage analyses make this assumption indirectly through the use of static prices that do not take into consideration the battery state. In other circumstances, production cost model is used in storage analysis, which minimizes the costs through optimizing the generator and storage dispatch. The arbitrage values can be exaggerated due to the static price assumption as it is possible for storage effect to neglect price differences. Therefore, when the operator maximizes the storage profit, storage could be underused compared with the ideal welfare. This suggests that the structure of contracts could be renewed to efficiently use the social value of a storage operation.

Another field of interest is the ability of storage to facilitate renewable energies market penetration. The main weakness of renewable energies is curtailment, which have a direct impact on the revenues obtained by these latter. This usually occurs because of constraints related to the operation and transmission. For example, in 2010, more than 8% wind potential generation was curtailed in Texas and 3.5% in the United States as a whole. Using storage devices could minimize the constraints related to this issue. In literature, the diverse interactions between renewable energies and storage systems are usually examined. Nonetheless, the number of detailed studies using utility-grade simulation tools to analyze the effects of renewable use on storage value is very limited. Coupling directly storage with a renewable can cause many inefficiencies, whereas there is a possibility to balance the variability using other sources that may include renewables. There is a growing necessity to delimit the role of storage in the integration of renewables and model the benefits of these services to accurately evaluate ES in the whole energy system.

Regulatory Treatment

In the recent years, the electricity industry has known a big change as it has shifted from the vertically integrated utility traditional model to restructured markets. In a traditional regulated market, "prudent" and reliable generation, transmission, and distribution investments are rate based by utilities. In a similar manner, storage systems are also rate based if they are shown to be the least-cost alternative to offer a service. However, this is difficult to do in reality because of the inability of the standard capacity expansion tools to accurately value ES. Some of the electricity services can be offered through competitive markets, whereas others cannot, which gave birth to the hybrid market design. An asset is classified traditionally in restructured markets as generation, transmission, or distribution. Investment in storage presents challenges because of its hybrid operation. Whether investment costs are recovered through the market or rate based is dependent on this classification, and this creates barriers for ESSs because they can have all of these characteristics. The absence of a regulatory definition of ES has resulted in its classification as a generation asset [38]. The value of ES is underestimated by this treatment because it neglects its capability to perform multiple services.

Techniques used to evaluate ES are becoming available, but they are still not commonly used. In addition, ancillary services are not recognized completely or well defined by regulators. Treatment of emissions is another aspect that should be considered in the regulatory environment. In the past, regulators and utilities concentrated on environmental compliance. Regulation associated with CO_2 emissions is still at an early phase. Carbon pricing has been approved by some utilities while considering competing resource expansion decisions. Example of explicit expressions of carbon cost that may be considered is the British Columbia's carbon tax as well as California's cap and trade system. The regulation of carbon emissions has been tackled also by the environmental protection agency.

The path toward reduced carbon emissions is well defined, and some utilities are becoming more worried about the risk of future strict carbon emission regulation. A higher share of renewable energy accompanied with ES would be driven by expected reduction in fossil fuels use and cost of renewable energy resources.

The use of ESSs as a substitute to the expansion of transmission lines is one of the most interesting value of ES. In the United States, regional transmission planning entities have been formed following FERC's Order 1000, which recognized the requirement for a better coordination among generation planning and transmission. However, those efforts are still at an early phase. Better communication between transmission planning functions and generation expansion is necessary or perhaps should be mandated by regulations.

Different and Altered Rules Within Regional Markets

Regulations, policies, stakeholders, and rules differ from one energy market to another. Each operator has its own characteristics such as requirement of capacity, restricted participant, and compensation. The treatment of ES depends on the regional independent system operator. Because of this different treatment, developers encounter diverse challenges while operating in multiple markets. Operators will be required in the near future to changes regulations and rules of ES to properly evaluate ES offering multiple services.

Development Risk

ES development risk is increased by the uncertainty and inability to reflect the real value of these systems. Potential new market entrants are also discouraged by this risk. Reasonably more complicated valuation approaches are necessary to reveal ES prudency to regulators would discourage utility planners from using prudent declaration. The dependence on market prices in emerging markets for new services increases the uncertainty for recovering capital-intensive costs. The highly variable markets associated with arbitrage value for ancillary services and energy shaping adds more uncertainty to the recovery of capital cost. ES value depends mainly on the provision of energy arbitrage and ancillary services, as well as the reduction of greenhouse gas emissions. The reliance on these revenue streams adds more risk to ES projects. In addition, the actual regulatory treatment of storage adds more risk to the deployment of new ESS. This latter suffers from a lack of incentives in regulated markets.

Although regulatory risk and market are important challenges, technology risks are also considered a significant issue for storage deployment. A number of storage technologies appear to be technologically viable, but they suffer from developers' reluctance, which has led to a delay in the development of these systems. For example, the construction of a second CAES plant in the United States has not yet occurred even though the operation of McIntosh CAES plant is successful and reliable. One of the main barriers opposing the construction of this is the geologic requirement.

The development risk is affected by the aforementioned uncertainties. These also discourage third-party development. Therefore, these risks should be addressed. In addition, their share between power system participants should be known. It is important to spread these risks among the potential beneficiaries and reduce them where possible. To encourage an accurate development, policies should be implemented to manage risks. Policies used to overcome obstacles or encourage the development of ES are examples of mechanisms for risk handling.

The development of small but significant scale of ES inspects and provides a market over which project proposals and systems may compete. The development risk would be left open with a loss of opportunities and technologies undeveloped if some level of development has not been tried. However, there should not be an overdevelopment of ES technologies at the expense of other potential alternatives such as demand management and renewable generation control strategies. A balance among the different existing strategies should be adopted, particularly with respect to system-specific incentives, to allow the implementation of competing alternatives.

Examples of mandates that have an impact on minimizing ES development risk are the California's and Puerto Rico Energy Authority's policies [39]. The development of cost-effective ES is mandated by California. In addition, the policy of Puerto Rico Energy Authority states that the deployment of new PV resources should be complemented by ES to maintain frequency control and system reliability. These two policies have an aim to evaluate the need for ES.

Public Attitude and Industry Acceptance

Public attitude is another important aspect that should be considered for the deployment of ES systems. Investment in these may be affected by the opinion of people toward energy systems. This is an important barrier that is sometimes ignored. The integration of ES into the society and people's life may not be always desirable. A boarder adoption of ESSs is faced also by a key challenge associated with utility industry acceptance. Requirements set by regulators foster more caution in implementing new technologies.

Grid Fees and Taxation

Some countries impose a double fee to energy transferred from and to ESSs due to the lack of a clear ES definition. That is, the owner of these facilities should pay grid fees for both the storage and the discharging of energy. In other European counties, grid fees are paid for only the generated energy. Taxes are being imposed on stored and self-consumed energy in other regions such as Spain. Because of the several benefits provided by ES to the power system, energy exchange taxation should be lowered.

Subsidies for Energy Storage Deployment

Even though ES plays a significant role in the electricity grid through the provision of services, it is challenging to justify its use due to the high capital cost it requires. Other alternative cheaper solutions exist, such as the reinforcement of the electric network and the use of flexible generation systems. To compete with these latter options, ES necessitates the use of subsidies. ES deployment is not supported by subsidies in some countries. This is considered an important barrier as it reduces the number of investors. The use of incentives will lead to the creation and growth of ES markets, which will in turn results in the reduction of costs. The development of innovative ESSs is encouraged by some energy markets in some states such as California. This has enacted a law in which utilities were required to provide procurement targets of ES for the year of 2020. Each utility and application of the electric network including the end user, the transmission, and the distribution have been set a particular ES target. This has resulted in an exceptional deployment rate of ES within that market. The risk of investing in ES decreases significantly with the use of such initiatives. Analogous obligations have been implemented lately in other countries such as France and China.

Lack of Standardized Controls and Interfaces

The lack of standardized interfaces and controls is considered a significant obstacle to industry acceptance of ES. This is considered a technical challenge that may result in implications associated with policy. Strategies to work across systems and utilities to develop standardized interfaces are anticipated. National Institute of Standards and Technology (NIST) is working on standardization. This is also tackled by the International Electro-Technical Commission (IEC).

There exist various IEC standards for mature ESSs such as PHS, Li-ion, lead-acid batteries. These standards include technical features, integration, and testing. There are only few standards for other storage options. Up to now, there is no storage system—independent standard for the integration of these systems into a stand-alone or utility grid. A standard dealing with any type of rechargeable batteries is planned.

Standardization of ESS includes the following:
- Terminology
- Characterization of ESS components for technical evaluation (power, capacity, lifetime, size, discharge time, etc.)
- Communication (security and protocols)
- Requirements for grid interconnection (synchronization, power quality, frequency-voltage tolerances, and metering)
- Mechanical and electrical safety
- System testing
- Implementation guidelines

Control standards would likely be addressed by utilities given the large diversity of ESS exciting in the marketplace, with a variety of parameters and different control considerations. Standardization is currently an issue that is open for discussion and research as there is a lack of maturity of ES in marketplace.

Standardizing communication protocols require a combination of efforts from several parties, which include grid operators, manufacturers, regulatory bodies, and utilities. However, it is still unknown whether the best solution for ES is a single communication standard or a new refined approach is required given the variety of the existing utility scale and the diversity of ES technologies.

Compared with other power system components, ES may face other barriers due to its different nature. To evaluate realistic interactions between ESSs and the electric grid, procedures should be developed by the transmission provider. An assessment of ES interconnection request should be performed as these systems consume and generate energy during low and peak periods. More realistic operating regimes should be recognized by standardized guidelines, which require ESSs to generate power only when other sources are limited and to consume power only when energy demand is low.

The technical requirements for connecting ES to the transmission grid should be defined by grid codes. The gird stability and operation is significantly affected by the growing penetration of renewable energy and ESSs; that is why a set of rules defining the standards, responsibilities, and requirement for every facility connected to the electric utility grid is required.

Energy Storage Network Cost

Although the funding of research in ES and the exploitation of different RES has increased significantly, electric ES is not witnessing a wide deployment in the electricity market. This is due to the fact that ESS operators receive only low credits compared with the important services they offer, particularly in restructured markets.

Reducing the peak loads in the electric grid highly depends on the support provided by ES. However, this is not the case in some countries such as the United Kingdom due to the high costs of "transmission network use of system" and "distribution use of system." These two costs have to be paid by ES operators in these countries when providing or storing energy. In case energy generation is less than 100 MW, the operator is exempt from paying the transmission network use of system but has to pay the distribution use of system. These tariffs generate a flow of revenue, which

could make ES valuable. The management system, which is in charge of controlling the operation of ES in the grid and in the electricity market, is a key for generating revenues and optimizing ES.

ES regulations are not imposed in a similar manner for each country by the European organizations. Some countries such as Spain, Portugal, Italy, Poland, Czech Republic, and Lithuania do not require the storage plants to pay charges; however, some others such as Greece, Belgium, and Austria have set charges on storage plants operating in both modes. The absence of specific European ES regulations leads each member country to define its policies and charges in the electric system. Because of this, member countries that do not impose such fees may invest in exporting electricity to those that impose grid charges. This would result in the formation of regional condensation of storage stations or plants in some places rather than others, which might lead to inefficiencies.

The realization of charging regimes in countries that consider ES as an energy generation and impose fees on them highly depends on the redefinition of ES as an independent asset in the energy market. A solution to this issue could be the implementation of grid connection fees based on the storage size, location, and type of usage. This could be complicated because the effect of these different variables on the storage value still needs further research.

SMART GRID AND ENERGY STORAGE

To ensure an efficient energy transition and provide good quality services, it is important to invest in a number of low-carbon and innovative energy systems. The accelerating R&D in renewable energy field is shedding light on the importance of ES deployment in renewable energy integration and smart grid. Nonetheless, the full potential of ESSs is not well depicted because their benefits with regard to smart grids are not well conceptualized. Therefore, for smart grid technology to develop and become more efficient and sustainable, the added value of ESSs needs to be understood. The deployment of ESSs, as part of the electricity grid, has been significantly hindered and limited due to their high cost as well as the undermining value, efficiency, and functionality of some systems. Moreover, the true potential of these systems is not well captured as there is a tendency to associate their application solely to load balancing. The recent and spurring ongoing interest in smart grid designs presents a new and exceptional opportunity to challenge the traditional model. Smart grid design planners have to take into consideration renewable energy generation and ES to maximize the efficiency of the electric grid and minimize carbon emissions.

Interest in ESSs has increased due to the transformation of conventional grids into smart grids. This later would lead to an improvement of renewable energy integration and an increase in the use of DG. In addition to this raising interest, the development of smart grid systems started focusing more on integrating intelligent and effective services of numerous assets to efficiently provide sustainable and economic energy. The relationship between both systems is becoming reciprocally advantageous. For instance, smart grid information technology enables ES to achieve its full potential. The ability to access and use real-time data related to power supply, demand, and quality as well as pricing allows ES to become valuable all along the power supply chain.

ESSs are also essential to smart grid efficient operation as they enable it to meet its objectives. Smart grids offer various advantages to the utility companies, such us minimizing the expensive interruptions, improving reliability, lowering distribution losses, increasing power delivery efficiency, and deferring capital expenditure on expensive generation and transmission resources. All the aforementioned advantages can be met through the use of ES as part of the electricity grid system. ESS has the ability to eliminate the necessity of using extra generation through load balancing. Moreover, investment on small-size ESS defers the required T&D upgrade to satisfy load growth. In addition, its use in substations can lower the congestion fees related to transmission in deregulated markets. Finally, the intermittency of RES is improved by the use of ES in a smart grid. ES is essential in improving the dispatching of renewable generation at the request of the grid operator. It offers different opportunities to counterbalance the demand and supply mismatch, as it also enables the distribution system to function in an economic, environmental, and more importantly efficient way. ES use coupled with smart grid technologies provides important benefits to residential consumers as well. These advantages include cost savings, thanks to the use of efficient systems, the ability to manage load during peak demand hours, and the use of affordable distributed RES. Similarly, smart grid also offers the possibility to control and dispatch storage units' loads. This makes distributed energy systems such as PV and wind energy technologies useful and valuable to the electric grid.

The possibility to integrate and deploy a large range of technologies is one of the most important advantages of smart grid. For example, flywheel can be used to overcome problems related to intermittency. This has a direct impact on enhancing the reliability and the power quality of the grid. Smart grid offers great services to ES and can help decrease losses related to standby efficiency.

The increasing use of renewable energy systems might cause some grid issues related to supply and demand. These can lead to the damage of grid electronic equipment and to the frequent occurrence of power outages. The demand flexibility can be improved by the use of ESSs with smart grid technology. These do not only enable the balancing of supply and demand but also increase the integration of renewable energy systems into the electric grid.

Even though ESSs provide a wide and versatile range of services to the grid, it cannot deliver these services continuously. Nonetheless, most necessary technical advantages can be delivered by smart grid while optimizing the obtained revenues. Therefore, to improve the potential of ES and maximize its value, it should be integrated into smart grid, thanks to the exchange of services as well as information. The assets of ES in the value chain highlight its importance in smart grid implementation. Nevertheless, ES capacity is insufficient to support the grid activities. For this reason, ESS policies need to be enhanced to promote and improve its deployment in the future.

RESEARCH AND DEVELOPMENT

The development of ES in the scope of smart grid needs to be encouraged by the performance of more research and development about new promising ESSs and the improvement of already existing ones. On the other hand, forecasts predict that ES costs are going to be reduced because of the developments in the field. Research, development, and demonstration offer a number of benefits to ES, including the reduction of their initial investment costs, the improvement of revenues obtained through the provision of services in different markets, and the demonstration of their viability. Funding and investment in R&D in the field of ES have many different extents to which projects could be attributed. Some examples could be the development of the newly arising technologies, the improvement of the already existing ones, or the enhancement of their design and capacities.

One of the most common barriers that face the development of new ES technologies is the technology risk. The development of various ES technologies has been delayed due to the reluctance of developers. The feasibility of new technologies can be shown through pilot and demonstration projects, which do not only show the viability of these systems but also reduce their development risk. In addition, pilot projects attract investors to invest in scaled-up systems used in large-scale applications. There are a number of ES technologies that have proved their feasibility through small-scale pilot projects; these include flywheel, batteries, and CAES. Examples of batteries include lithium-ion, vanadium redox, and sodium-sulfur batteries. Usually, the same pattern must be maintained for newer technologies starting from government-supported demonstration or pilot projects to private investors. Even though some technologies follow this pattern, they are limited when it comes to making them applicable for large-scale storage.

To improve of the existing ES technologies, more research funds are needed. The US department of energy has reported that research, development, and demonstration of ES technologies must be about reducing their costs while increasing their efficiencies. In addition, the performance of other systems needs to be improved. For example, flywheels are characterized by a short dispatch time, which limits their provision of services and thus reduces their incomes. R&D is the ultimate solution to enhancing the physical and technical capabilities of newly developed or already existing ES technologies. This will enable them to represent a bigger portion of the energy market.

CONCLUSION

ES is considered a key solution to improve the way conventional grids are structured and operated. However, these systems face a large range of barriers, including the incomplete and inaccurate valuation of ES revenues, the regulatory treatment, the lack of requirement and standardization, as well as the development risk. These challenges hinder their widespread deployment in the electric grid system. This has an impact on the potential applications as well as revenues that could be obtained through the provision of grid services. All the previously stated barriers are due to the lack of status definition of ES. The value of ES is underestimated in the restructured and the traditional markets due to the current regulatory structure. The full potential value of ES is not obtained because of the rate-based system used. This restricts the revenues obtained from performing multiple services, such as regulation services. Not all the services provided by ES are priced in market-based systems. T&D relief is an example of these services.

Restraining the role or type of ES has a direct impact on the full potential value of storage. There are some key market functions that are not effectively provided by other systems, such as power stability and energy management. The traditional way in which ES role is perceived as supplementary asset to the grid generates regulatory ambiguities. Regulators are facing some difficulties in pricing ES services and integrating their advantages as part of the grid components. This is mainly because of its polyvalence as it can provide a large variety of services ranging from generation to distribution.

Grid assets can be classified into three main categories, which include generation, transmission, and distribution assets. ES facilities are classified into only one of the three classes. This constrains and limits their classification as a supplementary asset class in the electric grid. Therefore, the core role of storage is stymied making its regulation difficult to establish. In addition, its pricing policy and tariff regulation are also tricky to define. The ambiguous classification of storage in the electric grid system does not allow its valuation to reach its full potential in a market place. Markets are not regulated and designed in a way to integrate the kind of services that are not part of an existing value chain. The economic viability of ES is hindered by the fact that it does not have a distinct status in the value chain. This makes it hard to regulators to establish rules and regulations enabling the expansion of its revenue streams. Therefore, ES needs to have its own asset class to be distinguished from other existing asset classes.

Although technical and economic issues are considered as the main challenges facing ES, market and policy issues represent important barriers to the development of ES. The lack of clear requirements for ES operation presents challenges. Moreover, the current markets are not well developed for some grid-ancillary services. In addition, they do not identify emission savings that may result due to the use of ESSs. Therefore, the real value of ES is not reflected and may be insufficient to justify its investment.

The difficulty of energy markets to value all the different benefits of ESSs, and the high investment risks put forward a need for policy intervention. Several policies have been executed to support other energy production systems for several purposes, such as feed-in tariffs, regulatory requirements, production tax credits, and others. Similar policies are necessary to obtain the complete economic and environmental advantages available through the use of ESSs.

REFERENCES

[1] Tiwari GN, Mishra RK. Advanced renewable energy sources. Cambridge: Royal Society of Chemistry; 2012.

[2] World Energy Council. World energy resources: hydropower. 2016. https://www.worldenergy.org/wp-content/uploads/2017/03/WEResources_Hydropower_2016.pdf.

[3] Electrical Power Research Institute. Handbook of energy storage for transmission and distribution applications. 2003. Washington, DC.

[4] Moe E, Midford P. The political economy of renewable energy and energy security: common challenges and national responses in Japan, China and Northern Europe. UK, London: Palgrave Macmillan; 2014.

[5] DOE Global Energy Storage Database. http://www.energystorageexchange.org/.

[6] Directorate general for internal policy, energy storage: which market designs and regulatory incentives are needed?. 2015. Brussels, Belgium, http://www.europarl.europa.eu/RegData/etudes/STUD/2015/563469/IPOL_STU(2015)563469_EN.pdf.

[7] Hane GJ. Government-promoted collective research and development in Japan: analyses of the organization through case studies. 1990. United States.

[8] Ministry of Economy, Trade and Industry (METI), establishment of storage battery strategy project team. 2012.

[9] Ministry of Economy, Trade and Industry (METI), fourth strategic energy plan. 2014.

[10] Tomita T. Policies and regulations for electricity storage in Japan. The Institute of Energy Economics, Japan (IEEJ), New and Renewable Energy and International Cooperation Unit; 2014.

[11] Yano Research Institute Ltd. Stationary ESS (energy storage system) market in Japan: key research findings 2015. 2015. Tokyo, Japan.

[12] Federal Energy Regulatory Commission (FEREC). Pumped storage projects. 2018. https://www.ferc.gov/industries/hydropower/gen-info/licensing/pump-storage.asp.

[13] U.S. Energy Information Administration (EIA). Electric power monthly with data for March 2018. 2018.

[14] U.S. Department of Energy. Grid energy storage. 2013. https://www.energy.gov/sites/prod/files/2014/09/f18/Grid%20Energy%20Storage%20December%202013.pdf.

[15] U.S. Department of Energy. Overview of development and deployment of codes, standards and regulations affecting ESS safety in the US. 2014.

[16] U.S. Department of Energy. Energy storage program planning document. 2011.

[17] The U.S. Department of energy's advanced research projects agency-energy. Washington, DC: FY 2014 Congressional Budget; 2014.

[18] Department of energy announces funding to support long-duration energy storage. Advanced Research Projects Agency-Energy (ARPA-E); 2018. https://arpa-e.energy.gov/?q=news-item/department-energy-announces-funding-support-long-duration-energy-storage.

[19] The U.S. Department of Energy's advanced research projects agency-energy, Duration Addition to electricitY Storage (DAYS). 2018.

[20] Federal Ministry of Economics and Technology (BMWi). Energy concept for an environmentally sound, reliable, and affordable energy supply. 2010. https://www.osce.org/eea/101047?download=true.

[21] Federal Ministry for the Environment. (BMU), Marktanreizprogramm für Erneuerbare Energien. Teil KfW: Förderung mit Investitionszuschüssen; 2013.

[22] China Energy Storage Alliance (CNESA). Energy storage industry white paper. 2018. https://static1.squarespace.com/static/55826ab6e4b0a6d2b0f53e3d/t/5b4bfa9f575d1fa91f3c8b35/1531706030137/CNESA+White+Paper+2018+%28English+Summary+Version%29.pdf.

[23] Document n° 1701 of China National Development and Reform Commission (NDRC) and the National Energy Commission (NEC), guidance on promoting energy storage technology and industrial development. September 22, 2017.

[24] China Energy Storage Alliance (CNESA). Large-scale energy storage projects around the country suggest 2018 will see a surge in energy storage growth. 2018. http://en.cnesa.org/latest-news/2018/8/15/large-scale-energy-storage-projects-around-the-country-suggest-2018-will-see-a-surge-in-energy-storage-growth.

[25] Decision n° 746-15 of the Moroccan Institute for Standardization. March 5, 2015 (Government's official gazette n°6348).

[26] International Electrotechnical Commission (IEC). IEC 61427-1:2013. 2013. https://webstore.iec.ch/publication/5449.

[27] Dahir n° 1-10-16 dated 11 February 2010, in Government's official gazette n° 5822. March 18, 2010.

[28] Dahir n° 1-16-3, in Government's official gazette n° 6436. 4 February 2016.

[29] http://rabatinvest.ma/index.php?option=com_content&view=article&id=4820:electricite-propre–le-marche-libre-arrive&catid=2:actualite&Itemid=174&lang=fr&listing=1.

[30] Ministère de l'énergie, des mines, de l'eau et de l'environnement, Note de présentation du projet de loi relative à la régulation du secteur de l'électricité, Rabat, Morocco. URL, http://www.sgg.gov.ma/portals/0/AvantProjet/129/Avp_loi_48.15_Fr.pdf.

[31] Short TA. Electrical power distribution handbook. 2004. https://goodboygunawan.files.wordpress.com/2010/03/electric-power-distribution-handbook.pdf.

[32] SBC. SBC energy institute analysis based on US DOE energy storage program planning document. 2011.

[33] Sioshansi, R., Denholm, P.,Jenkin T., Market and policy barriers to deployment of energy storage.

[34] Walawalkar R, Apt J, Mancini R. Economics of electric energy storage for energy arbitrage and regulation in New York. Energy Policy April 2007;35:2558–68.

[35] Hogan WW. Contract networks for electric power transmission. J Regul Econ September 1992;4:211–42.

[36] Drury E, Denholm P, Sioshansi R, August. The value of compressed air energy storage in energy and reserve markets. Energy 2011;36:4959—73.

[37] Mokrian P, Stephen M. A stochastic programming framework for the valuation of electricity storage. In: 26th USAEE/IAEE North American Conference. International Association for Energy Economics, Ann Arbor, MI; September 24—27, 2006.

[38] Giorgio C.G., P.E. Dodds, Jonathan R., Regulatory barriers to energy storage deployment: the UK perspective. Available at: http://www.restless.org.uk/documents/briefing-paper-1.

[39] Puerto Rican Power Authority press release. December 12, 2013. http://www.prepa.com/spanish.asp?url=http://www.aeepr.com/noticias.ASP.

CHAPTER 7

Conclusion

The question of how to achieve a sustainable power future is considered one of the main issues faced by modern society. An answer to this question could be the use of renewable energy sources. However, a significant implementation of these requires the deployment of energy storage systems. This book has explored the development of new novel energy storage system named gravity energy storage (GES). This technology is considered as an alternative to pumped hydroenergy storage. It has looked at this storage system from both technical and economic perspectives. In addition, it has investigated the feasibility, the profitability, and the performance of such a system with an aim of providing the reader insight about the development of this innovative technology.

GES is still in the demonstration and research phase as there is yet no operational life project to ascertain the system characteristics, applicability, and viability. More research is therefore necessary to investigate the realization of such storage technology. Major conclusions of the book are presented in this section.

The different conducted studies have shown that there is certainly potential in the investigated storage concept. A number of aspects were investigated and explained to achieve an overall study of the system. Although still some facets need further investigation, the following key conclusions were reached:

An increasing need of energy storage is expected in the next upcoming years. There exist several energy storage options. The most satisfactory and dominated method is pumped hydroenergy storage with more than 98% of the total installed capacity all over the world. However, this technology has an important environmental impact and needs large height difference between its two reservoirs. This latter issue prevents its development in low-lying countries. Other storage systems have different potentials and barriers. Compressed air energy storage is a profitable system. However, it does not provide pure and sustainable energy storage, as it uses natural gas. Batteries are also profitable, but they suffer from large environmental and safety challenges. In addition, they are not yet implemented on large scale. They require high costs, which will need to be dramatically decreased to become widespread.

Capacitors, superconducting magnetic energy storage, and hydrogen storage demonstrate great promises but are not used in large-scale energy storage applications. Currently, there is no storage technology able to compete with the profitability of pumped hydrostorage (PHS). Therefore, based on the principle and the main components of the conventional PHS, alternative storage options are being proposed and developed such as GES. This is a concept based on the idea of using the proven PHS technology in an innovative manner.

Chapter 2 investigates the design, construction, and sizing of GES. The proposed design methodology had an objective to maximize the storage capacity and minimize the construction cost while avoiding system failure. The obtained results demonstrate that to obtain an economic design for a specific storage capacity, the container height has to be increased while decreasing the size of its diameter. In addition, an investigation of the storage construction material was also conducted. It was found that iron ore is an optimal material that could be used to construct the piston because of its high density and lower cost compared with the investigated materials. Concerning the construction of the container, reinforced concrete material would be the best candidate. This chapter presented a methodology to optimally size and operate storage systems when coupled to wind farms. The aim of the proposed approach is to maximize the owner profit. The outcomes of this study show that a proper sizing of energy storage can increase the revenues obtained by optimally charging and discharging energy while taking advantage of energy price fluctuation. Toward the improvement of this energy storage technology, a novel concept has been proposed. This system is based on the addition of compressed air to GES. In Chapter 2, this new concept is introduced and studied. Modeling approaches are presented to determine the system feasibility. The outputs of this study reveal that the system storage capacity is significantly improved with the use of compressed air. Therefore, an interesting storage capacity could be achieved with the addition of compressed air to GES.

The economics of energy storage has been discussed in Chapter 3. The levelized cost of energy (LCOE)

Gravity Energy Storage. https://doi.org/10.1016/B978-0-12-816717-5.00007-4

approach was used to examine the viability of GES and compare it with other storage alternatives. It was found that the system has an interesting LCOE. Therefore, it is economically feasible to develop such technology. In addition, this chapter presented an explanation of electricity market structure. A model has been proposed to identify the different revenues available to energy storage participating in arbitrage and ancillary services within multiple markets including day-ahead and real-time energy markets. The outcomes of this study show that operating in regulation market is likely to provide the most potential benefits. However, gravity storage has to provide multiple grid services to achieve a positive net present value (NPV). The concept of the GES is considered a low-risk solution, as it uses the principle of the proven PHS technology in an innovating way. As structural components, only a container, return pipe, sealed piston, and pump turbine-station are required. The key risks of further commitment to the construction of this storage system are mainly based on uncertainties concerning the difficulty of excavating deep shafts and the use of large sealing system. In addition, the development of such energy storage project is affected by external risks associated with the economic, financial, and political aspects of the project.

Chapter 4, built on the work performed in Chapter 3, investigates the profitability of energy storage in small- and large-scale applications. This has been performed through the use of different interconnection scenarios. A cost-benefit analysis demonstrates that GES is not viable for residential small applications except if it is installed for transmission and distribution (T&D) upgrade deferral application. However, the system generates a positive NPV when used for large-scale application.

A sensitivity analysis has been conducted to delve into the impact of changing critical variables on the project profitability. The results reveal that this system is not economically viable if faced by an 18% increase of the project investment cost. Moreover, the project NPV is significantly affected by a decrease in potential revenues generated from performing arbitrage and T&D deferral. Finally, an increase in the discount rate negatively impacts the viability of this storage technology. It has been observed that a 3% increase of this would result in a negative NPV. Finally, a feasibility study investigating the incorporation of GES in building has been performed in this chapter. The obtained results demonstrate that the system is not economically competitive with other energy storage options used in building applications. This is due to the low LCOE it achieves in this latter application.

Chapter 5 presented dynamic modeling methods for GES. The behavior of this energy storage is studied to gain insight into the system performance. In addition, this study details the operation modeling of this storage system coupled with a photovoltaic energy plant. The obtained results show that the proposed model meets the load according to the presented management strategy. As for the hydraulic modeling, the proposed model enables the identification of critical parameters such as the system flow rate, volume, pressure, and the piston dynamics, as well as the system discharge time. The results obtained from the simulated case study were compared with experimental results of other researchers and evaluated. The % errors of the system discharge time and pressure are rather small. The proposed mathematical model has demonstrated its ability to accurately simulate the system dynamic response. Therefore, it is possible to model a simulated system with an aim of identifying its critical characteristics and behavior.

In addition to the economic and technical barriers faced by energy storage developers, there are also challenges associated with policies and markets, which lead to more reluctance toward the development of energy storage. This book ends up with a discussion about policy consideration and future prospects in Chapter 6. Interest in energy storage has been fueled by the increasing development of intermittent renewable energy generation. However, the deployment of energy storage systems has been hampered by their relatively high capital costs compared with traditional generating resources. In addition, energy storage development faces other barriers, which suggest the intervention of policies. It is still complex to determine the real value of energy storage, which results in an incapability of recognizing the system benefits.

With the growing share of renewable energy sources, there will be an increasing demand for energy storage services. Even if energy storage is not the only provider of these services, there is actually a significant emphasis on their development. This later is fostered by policies that reduce challenges and makes incentives available for developing energy storage. Standardized integration of energy with utility system is also needed and merits development. In addition, energy storage may become economically sound if policies are adopted. These policies and standards would attract investors and provide support to the implementation of promising energy storage systems that are still in development phase.

Even if energy storage is considered as a key system, the exciting regulations does not unable its development within the power system as a flexible solution.

Various uncertainties and issues are created with the current treatment of energy storage as a generation source. Finding an optimal solution to these challenges could simplify its deployment. In addition, the impact of these barriers results in issues related to compensation, price signals, and delays. Numerous initiatives should be undertaken to deal with energy storage issues associated with revenue streams. Although various changes in some markets have been achieved such as fast reserve and frequency response services, further efforts are necessary. Additional research is required to gain insight into the different benefits that could be provided by energy storage. A number of policies are also needed to properly value energy storage systems based on their participation into the electric grid operation. These regulations and policies will allow energy storage to become an integral part of the power system and would remove its different challenges and barriers.

Index

Note: Page numbers followed by "f" indicate figures and "t" indicate tables.

Printed in the United States
By Bookmasters